STATE OF THE [illegible]
OF ENERGY EFFICIENCY:

Future Directions

5/24/91

to Tom and Penny

Great friends!

Ed

American Council for an Energy-Efficient Economy
Series on Energy Conservation and Energy Policy

Series Editor, Carl Blumstein

Energy Efficiency in Buildings: Progress and Promise
Financing Energy Conservation
Energy Efficiency: Perspectives on Individual Behavior
Electric Utility Planning and Regulation
Residential Indoor Air Quality and Energy Efficiency
Energy-Efficient Motor Systems: A Handbook on Technology, Program, and Policy Opportunities
State of the Art of Energy Efficiency: Future Directions

STATE OF THE ART OF ENERGY EFFICIENCY:

Future Directions

Edited by

EDWARD VINE

and

DRURY CRAWLEY

American Council for an Energy-Efficient Economy
Washington, D.C., and Berkeley, California

in cooperation with:
Universitywide Energy Research Group
University of California

1 9 9 1

State of the Art of Energy Efficiency: Future Directions

Published by the American Council for an Energy-Efficient Economy
1001 Connecticut Avenue, N.W., Suite 535, Washington, D.C. 20036.

Cover design by Wilsted & Taylor

Book design by Paula Morrison

Book typeset by Campaigne & Associates Typography

Printed in the United States of America by Edwards Brothers Incorporated

Library of Congress Cataloging-in-Publication Data

State of the art of energy-efficiency: future directions / edited by Edward Vine and Drury Crawley.
304 p. 23 cm. — (Series on energy conservation and energy policy)
Includes bibliographical references and index.
ISBN 0–918249–11–2: $24.50
1. Buildings—Energy conservation I. Vine, Edward L. II. Crawley, Drury, 1957– . III. American Council for an Energy-Efficient Economy. IV. University of California (System). Universitywide Energy Research Group. V. Series.
TJ163.5.B84S726 1991
696—dc20 91–12339
CIP

NOTICE

Acknowledgements

As noted previously, earlier versions of these papers were presented at the ACEEE 1990 Summer Study on Energy Efficiency in Buildings. As co-chairs of this conference, we realized that the content of these state-of-the-art papers should be presented to a wider audience, and with the encouragement of Carl Blumstein, president of ACEEE, we decided to publish this book. We are especially grateful to the reviewers of the conference papers and to the panel leaders who participated in the review process. We thank John Kesserling, Bill Krauss, Paul Centolella, Dave Grimsrud, Eric Hirst, Alan Destribats, Marty Kushler, Bobbi McKellar, Martha Hewett, Jeff Schlegel, Meg Fels, and Mimi Goldberg.

We are indebted to Stephen Frantz, who copyedited the book. This task involved careful reviews and revisions to reconcile the writing styles of the authors. Our thanks also to the production team at *Home Energy* magazine, Karina Lutz, managing editor, and Michelle Stevens, production manager.

Finally, we are grateful to the many organizations that supported the 1990 Summer Study and made the preparation of this book possible. Sponsors were Bonneville Power Administration, California Energy Commission, Electric Power Research Institute, Gas Research Institute, Lawrence Berkeley Laboratory, New England Power Service Company, New York State Energy Research and Development Authority, Ontario Hydro, Pacific Gas and Electric Company, Pacific Northwest Laboratory, San Diego Gas and Electric Company, Southern California Edison Company, U.S. Department of Energy, and Western Area Power Administration. Contributors were American Public Power Association, Brookhaven National Laboratory, California Institute for Energy Efficiency, Central Maine Power, The Fleming Group, Libbey Owens Ford

Company, National Energy Program Evaluation Conference, Niagara Mohawk Power Corporation, North Carolina Alternative Energy Corporation, Northeast Utilities, Oak Ridge National Laboratory, RCG/Hagler, Bailly, Inc., Rochester Gas and Electric Corporation, Seattle City Light, Southern California Gas Company, Texas Governor's Energy Management Center, and Wisconsin Power and Light Company.

Dru Crawley
Ed Vine

Preface

Oil Conservation or Oil Conflict: Alternatives to Our Nation's Dependence on Foreign Oil

The *ACEEE Summer Study on Energy Efficiency* took place the same month that Iraq invaded Kuwait. The United States responded to the Iraqi invasion by sending troops to Saudi Arabia. As a result, a conference devoted to saving oil, gas, and money suddenly attained a new relevance.

Many times during the Summer Study I thought, "Why go to war over oil in the Middle East when you can come to this conference (or read this book) and learn to save enough oil to cease our country's dependence on imports from the Gulf?"

How can we save all of that oil and gas (resources that are interchangeable in our industrial society)? This book addresses two interdependent issues that are at the root of our oil consumption woes:

1. How can consumers be encouraged to extend their time of purchase horizons and buy products that have lower life-cycle costs, rather than products for the lowest initial investment?
2. How can utilities be encouraged to diversify from merely selling energy to selling energy services? How can their profit rules be changed so that utilities investors earn more from selling efficiency than from selling raw energy?

State of the Art of Energy Efficiency: Future Directions offers valuable and timely answers to these two questions. The book also discusses how to reduce the twin threat: how our nation's present conflict over oil will affect our planet's global climate for generations to come. The book effectively answers all of these questions, and tells us how to save money while doing so. So read and enjoy.

A.H. Rosenfeld
Berkeley, CA
December 1990

Contents

Introduction

In the past ten years, energy conservation efforts have greatly expanded. Focused mostly on the buildings sector, which accounts for over 20% of current U.S. energy consumption, these efforts have been diverse. Energy-efficiency technologies and programs, resource planning and policy making, and data collection and analysis methodologies have all played an integral role in promoting energy efficiency. Given this diversity, there is a need for a useful and practical compilation of the state of the art in these areas. This book was developed to fill that need.

The book reviews current knowledge and addresses promising directions for designing and implementing research and programs that will promote energy efficiency in buildings. Topics include design and implementation of government and utility programs, appliance standards, collection and analysis of data on energy use in buildings, and integrated resource planning.

The chapters—written by leading researchers, program analysts, and policy makers—are based on papers prepared for the ACEEE 1990 Summer Study on Energy Efficiency in Buildings. The Summer Study is one of the premier conferences that attract international, interdisciplinary participants for the purpose of exchanging information and sharing ideas on how to improve the efficiency of energy use in buildings. For the first time in the history of the conference, eight researchers or research teams were requested to prepare state-of-the-art papers on key topics of importance to the conference. These selected topics, a subset of the broad range of topics in the energy efficiency of buildings, were felt to be especially critical to today's needs.

Issues Raised in This Volume

During the past few years, both the scope of resource planning at electric utilities and the diversity of participants in the planning process have grown dramatically. The scope of planning has expanded to consider energy-efficiency and load-management programs as resources, to take account of the environmental costs of electricity production, and to incorporate a variety of resource-selection criteria beyond electricity price. The mix of participants has expanded beyond utilities themselves to include regulatory commissions, nonutility energy experts, and customers. Similar changes are beginning to occur at gas utilities.

In chapter 1, Hirst et al. review a few of the key issues related to integrated resource planning (IRP), the latest form of utility resource planning. These issues include provision of financial incentives to utilities for successful implementation of integrated resource plans, incorporation of environmental factors in resource planning, bidding for demand and supply resources, development of guidelines for preparation and review of utility integrated resource plans, and resource planning for gas utilities. The authors also describe the need for greater efforts by the U.S. Department of Energy to encourage integrated resource planning.

In chapter 2, Ottinger discusses recent moves by 29 state regulatory commissions, Washington, D.C., the Northwest Power Planning Council, and the Bonneville Power Administration to consider environmental externalities in resource planning. The principal rationale for incorporating environmental costs into resource planning decisions and proceedings is that electric utility operations impose large damages to human health and the environment, damages not taken into account by traditional utility least-cost planning and resource selection procedures or by government pollution regulation. Traditional procedures have effectively valued at zero the residual environmental costs to society of utility operations.

Ottinger notes that although most regulatory commissions requiring consideration of environmental externality have left it to the utilities to compute the societal costs, a few commissions have either set those costs themselves or attached a proxy adder for polluting resources (or a bonus for nonpolluting ones). The commissions that have computed costs have based their computations on pollution control or mitigation costs rather than societal damage costs. Ottinger supports a methodology based primarily on assessing damage costs where adequate studies exist to permit quantification. In this chapter, he discusses the methodologies used for the measurement of damage costs and describes the approaches that have been and might be used for incorporating damage costs in integrated resource planning.

In chapter 3, Nadel examines the basic types of residential and commercial conservation programs that have been offered by electric utilities, including information, rebate, load management, performance contracting, comprehensive direct installation, and bidding programs. Discussions cover typical and exemplary programs and include information on key program attributes, such as participation rates, free riders, savings, and cost per unit of energy saved, as well as factors linked with program success. Nadel emphasizes programs and procedures that have achieved high participation rates and savings while being cost-effective to the sponsoring utility. While program experience to date teaches many important lessons about ways to structure and promote programs in order to achieve substantial energy and dollar savings, much remains to be learned if even half the technical potential for conservation improvements is to be realized.

As demand-side management (DSM) and IRP become more routine elements of the utility planning process, utilities' energy information needs have expanded well beyond sales and peak loads by customer class. An increasing number of utilities require credible estimates of both end-use load shapes and load shape changes associated with DSM options in order to assess the cost-effectiveness of alternative resource options. End-use energy data are notoriously difficult and expensive to collect, however, especially on an hourly basis.

In chapter 4, Eto et al. discuss the importance of end-use load shape data for IRP. Citing both load shape estimation studies and recent end-use metering projects, the authors summarize leading applications and review progress to date in obtaining load shape data. The historic origins and the state of the art in each topic area are discussed. Eto et al. conclude that current metering projects are producing valuable data broadly applicable to other service territories and that load shape estimation techniques have matured sufficiently to represent reliable tools for planning. Finally, the authors outline directions for future research by arguing for closer coordination between efforts to improve metering and those to improve load shape estimation techniques.

In chapter 5, Misuriello complements the previous chapter's perspective by describing the technology available to monitor building energy performance. He traces both current activities to develop the technology and the trends and needs that are stimulating these activities. Misuriello further explains what types of studies monitoring technologies make possible and what limitations we still face. This chapter emphasizes those areas of buildings research that have traditionally been outside the scope of utility load research. Misuriello concludes that widespread implementation of DSM and least-cost utility planning will most likely make field monitoring more critical in the areas of program planning, implementation, and evaluation.

In chapter 6, Schlegel et al. summarize the status of low-income weatherization programs, which are undergoing major changes in funding sources, program delivery systems, and level of technical sophistication. At the same time, the original purpose of these programs—easing the burden of energy costs on low-income households—is as important as ever. Government and utility programs are struggling with how to deliver effective services designed to provide significant energy savings in an era of dwindling resources and changing providers.

Schlegel et al. summarize information on such key issues as the evolution of weatherization technology, the disappointing level of energy savings achieved by the programs, the emergence of the utility industry as a funding source, client/customer education, the evolution of field personnel from estimator to auditor to technician, and the importance of ongoing program evaluation. They review past history and current practice and suggest future directions for low-income weatherization services, including both governmental and utility–private sector programs. This chapter should serve well both as a reference document for past activities and as an inspiration for moving ahead in the future to effectively address the major need for this service.

In chapter 7, Turiel et al. present the status and future of U.S. efficiency standards for residential electrical appliances. The authors review the history of the process for setting efficiency standards and suggest potential improvements. Issues related to the standards, including the roles of government and private industry, are discussed. Other policies for achieving energy savings from more efficient appliances—policies such as providing financial incentives for choosing more efficient appliances—are also examined.

The household sector has been the target of many energy-saving policies; however, many policies put in place in the late 1970s or early 1980s have expired. If the expiration of policies permits or encourages demand to increase, then a sizable component of past savings in energy and oil might be cancelled. Careful evaluation of the nature of savings in this sector since the early 1970s is, therefore, important in order to determine future growth in energy demand.

In chapter 8, Ketoff and Schipper examine trends in household energy use and the impacts of energy conservation on such use in the United States, Japan, and Western Europe from 1972 through 1987. Rather than relying on aggregate data, the authors measure conservation as the change in energy intensity by end use, defined as energy per unit of activity or output, and they estimate the amount of change in energy use due to conservation rather than to changes in other variables (income, housing stock, and family size and structure). The authors conclude that household energy use has dropped substantially since 1973, with most of the reduction occurring in space heating.

Final Comments

Given the breadth and quality of the information it presents, this book, we believe, will become an important resource for professionals in utilities, government, research institutions, and universities. In addition to helping define program priorities and funding levels, the book should stimulate readers to develop their own ideas for improving energy efficiency in buildings. The book should be viewed primarily as a reference (like an encyclopedia) rather than as a text (in which each chapter necessarily follows those that precede it), although we have tried to organize chapters in a logical sequence.

Chapter 1

Integrated Resource Planning for Electric and Gas Utilities

Eric Hirst, *Oak Ridge National Laboratory*
Charles Goldman, *Lawrence Berkeley Laboratory*
Mary Ellen Hopkins, *The Fleming Group*

During the past few years, the scope of resource planning at electric utilities and the variety of participants in the planning process have expanded dramatically. Planning now takes into account energy-efficiency and load-management programs considered as resources; the environmental costs of electricity production; and a variety of resource-selection criteria beyond electricity price. Participants in the planning process now include regulatory commissions, nonutility energy experts, and customers, as well as utilities themselves. Similar changes are beginning to occur at gas utilities.

Integrated resource planning (IRP) helps utilities and state regulatory commissions consistently assess a variety of demand and supply resources to cost-effectively meet customer energy-service needs. Key characteristics of this planning paradigm include (1) explicit consideration of energy-efficiency and load-management programs as alternatives to some power plants and new supplies of natural gas, (2) consideration of environmental as well as direct economic costs, (3) public participation, and (4) analysis of the uncertainties and risks posed by different resource portfolios and by external factors. IRP differs from traditional utility planning in several ways (Table 1-1, next page). Several of the sources referenced in this paper discuss IRP and its development (Cavanagh 1986; EPRI 1988; Gellings, Chamberlin, and Clinton 1987; Hirst 1988a and 1988b; and NARUC 1988).

Table 1-1. Differences between traditional utility planning and integrated resource planning.

Traditional Planning	Integrated Resource Planning
Uniformity of resources: utility-owned central-station power plants	Diversity of resources, including utility-owned power plants, purchases from other organizations, conservation and load-management programs, transmission and distribution improvements, and pricing
Planning internal to utility, primarily in system planning and financial planning departments	Planning spread among several departments within utility and often involves customers, public utility commission staff, and nonutility energy experts
All resources owned by utility	Some resources owned by other utilities, by small power producers, by independent power producers, and by customers
Resources selected primarily to minimize electricity prices and maintain system reliability	Resources selected on the basis of diverse criteria, including electricity prices, revenue requirements, energy-service costs, utility financial condition, risk reduction, fuel and technology diversity, environmental quality, and economic development

This chapter reviews recent progress in IRP and identifies the need for additional work. Key IRP issues facing utilities and public utility commissions (PUCs), discussed in this chapter, include

- Rewarding utilities for successfully implementing integrated resource plans and especially for acquiring demand-side management (DSM) resources
- Incorporating environmental costs
- Bidding for demand and supply resources
- Developing guidelines for preparation and review of utility resource plans
- Adapting IRP and DSM programs to the needs of gas utilities
- Developing federal policies to promote IRP and DSM

Many other issues related to IRP, but not discussed in this paper, are important. Such issues include alternative ways to organize planning within utilities; the uses of collaboration and other forms of involvement in planning by nonutility participants (Ellis 1989; Schweitzer, Yourstone, and Hirst 1990); the effects of competition and deregulation on utility planning; treatment of electricity pricing as a resource; fuel switching (primarily between electricity and gas); treatment of uncertainty in utility planning and decision making (Hobbs and Maheshwari 1990; Hirst and Schweitzer 1990); the appropriate economic tests for utility DSM programs; ways to measure the performance of DSM programs (Argonne National Laboratory 1989); development and use of improved data and planning models; and transfer of information among utilities and commissions (Goldman, Hirst, and Krause 1989).

Rewarding Utilities for Effectively Implementing Integrated Resource Plans

IRP treats energy-efficiency programs as resources that can substitute for some amount of generating capacity. Unfortunately, traditional regulation of utilities discourages this practice because "each kWh a utility sells . . . adds to earnings [and] each kWh saved or replaced with an energy efficiency measure . . . reduces utility profits" (Moskovitz 1989). This policy pits the interest of utility shareholders against that of utility customers. Realizing that effectively implementing integrated resource plans may hurt utility shareholders, the National Association of Regulatory Utility Commissioners (NARUC) passed a resolution urging PUCs to adopt ratemaking mechanisms that reimburse earnings lost through DSM programs and that thus encourage these programs (NARUC 1989).

Proposed reforms of PUC regulations would allow utilities to recover both operating costs of DSM programs and net lost revenue (revenue forgone because of reduced electricity use minus the reduction in operating costs). Reforms would also reward utility shareholders for exemplary delivery of DSM services. Cost recovery removes disincentives to DSM programs; rewarding shareholders adds positive incentives to run such programs. The underlying idea is that utilities should operate under regulatory practices that make it financially attractive for them to implement all aspects of their integrated resource plan, not just acquisition of supply resources (Moskovitz 1989; Wiel 1989).

Simple incentive methods have been in place for several years in a few states, including Washington and Wisconsin. In 1989, PUCs in several states, including Maine, Vermont, Massachusetts, New York, New Jersey, Minnesota, Nevada, California, and Washington began inquiries into the desirability and form of different incentive procedures. Progress has been most rapid in California, Massachusetts, New York, and Rhode Island. The New York Public Service

Commission approved proposals from several utilities to test incentive schemes,[1] and the Rhode Island and Massachusetts PUCs adopted incentives for New England ElectricSystem.

A successful regulatory system should decouple profits from sales (Moskovitz 1989). That is, utility earnings should not depend on the quantity of sales achieved. The Electric Revenue Adjustment Mechanism, used in California since 1982, allows utilities to use balancing accounts to adjust for the over- or under-collection of authorized base revenues (essentially all nonfuel costs) caused by discrepancies between actual and forecast sales of electricity (Marnay and Comnes 1990). Utilities then return these revenues to (or collect them from) customers the following year through an adjustment to the price of electricity. This mechanism breaks the link between sales and profits, thus eliminating a major disincentive to utility DSM programs.

Measures that go beyond removal of disincentives to reward utilities for effective programs include rate-of-return adjustments, shared savings, and bounty. The shared savings approach seeks to reward utilities for acquiring resources that deliver desired energy services at least cost. The two other approaches are less appealing because the utility incentive does not depend directly on the benefit provided by the utility DSM programs (Moskovitz 1989).

New England Electric System (Sergel 1989; Destribats, Lowell, and White 1990) proposed a shared-savings incentive scheme (Figure 1-1) in Rhode Island, New Hampshire, and Massachusetts. The proposal has two parts. The first incentive, intended to maximize the size of the company's DSM programs, equals 5% of the programs' total benefit, defined as the company's avoided costs (kWh saved × ¢/kWh + LW saved × $/LW). The second component, intended to reward program efficiency, equals 10% of the programs' net benefit, defined as avoided costs minus the costs of operating the programs. The Rhode Island PUC approved a modified version of this proposal. The Massachusetts Department of Public Utilities approved an incentive system with "a fixed payment for each kW and kWh saved that is verified through an after-the-fact evaluation and monitoring system."[2]

As part of the 1989 collaborative in California, Schultz (1990)

---

1. *Opinion and Order Approving Demand-Side Management Rate Incentives and Establishing Further Proceedings.* Opinion No. 89-29, September 1989, Albany.
2. *Investigation by the Department on Its Own Motion as to the Propriety of Rates and Charges . . . by Massachusetts Electric Company.* DPU 89-194 and DPU 89-195, 1989.

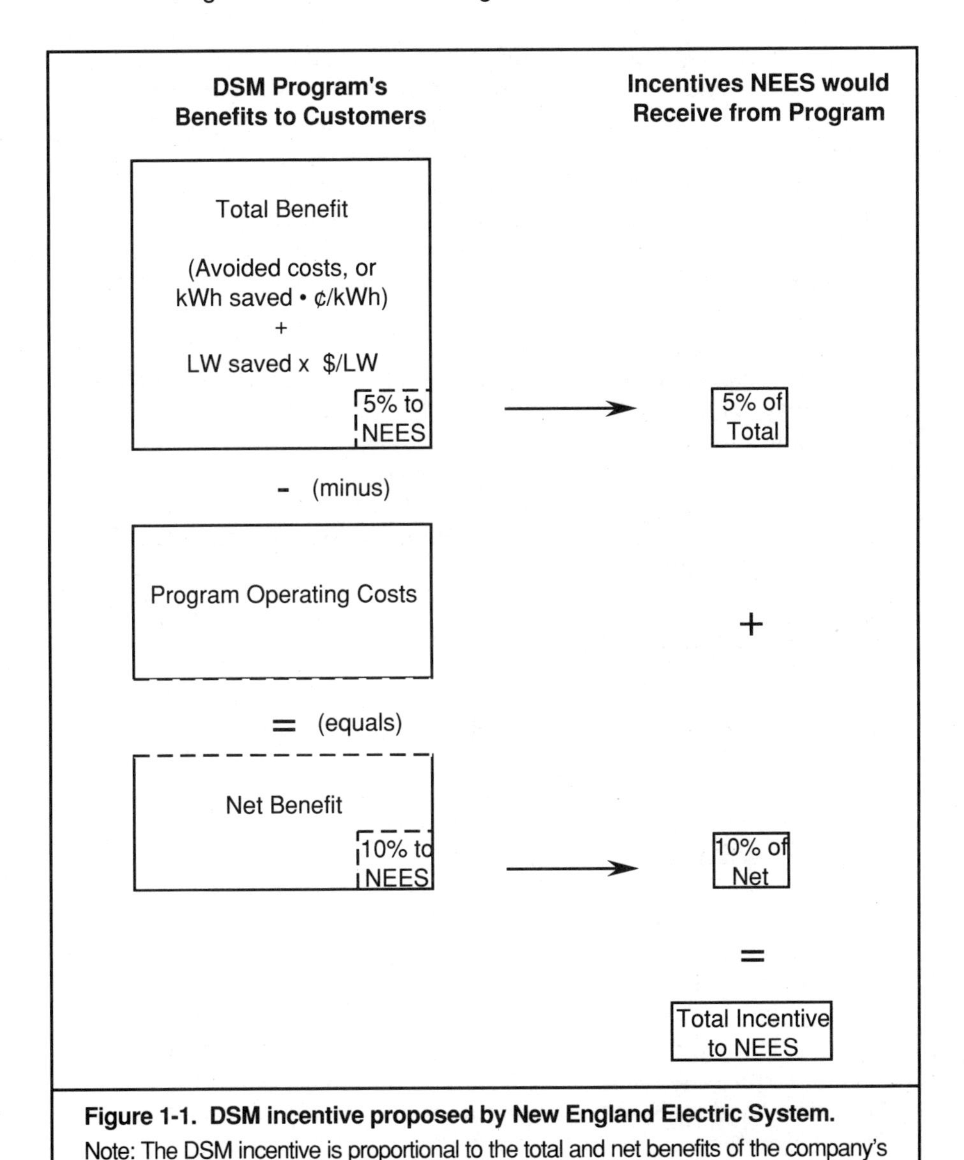

Figure 1-1. DSM incentive proposed by New England Electric System.
Note: The DSM incentive is proportional to the total and net benefits of the company's DSM programs.

examined alternative shared-savings proposals for utility DSM programs. His analysis focused on the purposes of the DSM programs and on the risk/reward relationships implicit in different incentive proposals. These proposals differed in their sensitivity to changes in total resource costs, utility costs, avoided costs, and electricity savings stimulated by the DSM programs. Schultz suggested that incentive mechanisms should seek to maximize the net benefit of DSM programs, set performance standards, and minimize program costs.

Reforming utility ratemaking is now an important part of IRP. Discussions among utilities, commissions, and others are under way in many states, and, in a few states, utilities are testing such schemes. Additional analyses of the pros and cons of different regulatory reforms within the context of the accounting rules used by individual states and utilities, and evaluations of the effects of these reforms, are needed. These evaluations should identify the effects of regulatory reform on the size and effectiveness of utility DSM programs and on the costs to utility customers. In addition, the energy savings and load reductions caused by utility DSM programs must be carefully and accurately measured (Argonne National Laboratory 1989), because these measurements determine the incentive payments to utilities.

Incorporating Environmental Costs

The environmental impacts of power plant operation have significant effects on society. For example, electricity production accounts for two-thirds of the SO_2, one-third of the NO_x, and one-third of the CO_2 emitted in the U.S. These airborne pollutants are linked to secondary effects such as acid rain and global warming, which reduce forest production, damage coastal land, and may produce other changes that affect society. These effects of electricity production are externalities, defined as any costs not reflected in the price paid by customers for electricity (Chernick and Caverhill 1989).

External effects cease to be externalities once their costs are paid by the entity responsible for their production and are reflected in the price charged for the product. For example, some environmental costs associated with electricity production are internalized by federal and state regulations that require mitigation of negative impacts on land and water use.

PUC Approaches

Partly in response to increased public concern about acid rain and global warming, several PUCs are insisting that environmental impacts be explicitly accounted for in utility resource planning and acquisitions. A recent survey found that 15 PUCs have procedures for considering environmental externalities in their resource planning and acquisition (Cohen et al. 1990). These approaches seek to influence the choice of resources, not to increase the costs of the resources that have already been chosen. However, the ultimate cost of electricity could rise if environmentally benign resources are more expensive than other alternatives.

Typically, these procedures internalize costs in one of three ways. The utility may treat environmental externalities quantitatively. For example, the Nevada Public Service Commission has broad discretion to "give preference to the measures . . . that provide the greatest economic and environmental benefits to the state." Under this approach, PUCs require utilities to consider environmental costs in resource planning but do not specify the methods to be used.

This approach has been used by Ontario Hydro (1990) (see Table 1-2). The utility selected environmental effects to use as criteria, which included the following:

- Impacts on land and water use
- Contribution to atmospheric emissions (SO_2, NO_x, CO_2, radionuclides, and trace elements), effluents (including thermal discharges, radionuclides, and uranium- and coal-mining effluents), and solid waste (coal ash, flue-gas desulphurization wastes, spent nuclear fuel, and uranium mine tailings)

Table 1-2. Environmental analysis process used by Ontario Hydro.

1. Develop criteria for environmental effects

   Resource use
   - Nonrenewable resource use: coal, oil, gas, and uranium
   - Water use
   - Land use

   Emissions, effluents, and wastes
   - Atmospheric emissions: SO_2, NO_x, CO_2, radionuclides, and trace elements
   - Aquatic effluents: thermal discharges, radionuclides, uranium mining effluents, and coal mining effluents
   - Solid wastes: coal ash, flue-gas desulphurization wastes, used nuclear fuel, low-level radioactive wastes, and uranium mine tailings
2. Evaluate the environmental implications of alternative plans
3. Consider mitigation/compensation to offset the potential environmental effects
4. Determine the environmental advantages and disadvantages of the alternative plans

- Socioeconomic effects such as regional employment and local community impacts

The utility then analyzed the environmental advantages and disadvantages of alternative demand and supply plans, taking into account the steps and costs required to mitigate the potential environmental effects of each plan.

A second approach uses a percentage adder that either increases the cost of supply resources or decreases the cost of DSM resources. For example, the Wisconsin Public Service Commission recently implemented a noncombustion credit of 15% for nonfossil and demand-side resources because these resources reduce pollution.[3]

A third approach quantifies the cost of the externality. This method is often used if the utility is developing a competitive resource procurement (bidding) process. For example, the New York Public Service Commission recently approved utility bidding proposals that assigned values to different levels of air and water emissions and land degradation of each bid, which can add up to 1.4¢/kWh.[4] Several of the New York utilities use this method to adjust the cost of each bid, while other utilities use a "point system," which weights environmental factors relative to the price factor in scoring competing supply and demand projects.

Alternative Methods of Quantifying Environmental Costs

In calculating external costs, planners must choose the proper method to use. Two basic approaches are used currently. The first approach calculates damage costs imposed on society by first tracing impacts of a generating technology through each step of its fuel cycle (emissions; transport of pollutants; the effects of these pollutants on plants, animals, people; and so forth), then estimating the extent of each impact and assigning a value to it. For example, SO_2 emissions can be linked to lost forest products, damage to buildings, and human respiratory problems. Quantifying the dollar cost of each effect is quite difficult, however: valuation is impossible for some environmental and resource damages, dose-response relationships are uncertain, valuing intangible costs such as those for recreation facilities and endangered species of wildlife is difficult, valuing human mortality is controversial, and some damages are

---

3. *Findings of Fact, Conclusion of Law and Order.* 05-EP-5, 6 April 1989, Madison.
4. *Opinion and Order Establishing Guidelines for Bidding Program.* Case 88-E241, Proceeding on motion of the Commission as to the guidelines for bidding to meet future electric capacity needs of Orange and Rockland Utilities, Inc., Decision 89-7, 12 April 1989, Albany.

very site specific (Chernick and Caverhill 1989; Ottinger et al. 1990).

A second approach bases the value of pollution reduction on the cost of controlling or mitigating the pollutants emitted by the generating technology (Chernick and Caverhill 1989). This approach assumes that the cost of controls reflects the price society is willing to pay to reduce the pollutant. This approach has limitations—for example, it cannot be applied to unregulated pollutants such as CO_2—but is nevertheless attractive because thorny issues of assigning dollar values to human life or valuing ecosystems have implicitly been dealt with by the legislators and regulatory bodies that formulated the pollution control standards. The disadvantage of using control costs to calculate environmental externality costs is that they typically bear little relation to the actual damages imposed on society by power plant emissions.

New England Electric System (NEES) has developed a hybrid approach it calls an issue-based rating and weighting index (Destribats, Lowell, and White 1990). NEES assigns to an environmental impact such as acid rain a weight based on a survey of experts. The company then assigns a rating from zero to four to every contributing factor, such as SO_2 and NO_x in the case of acid rain. The highest rated project—the one causing the most environmental degradation—receives a cost adder of 15%, and the costs of other projects are increased by a ratio of their score to the highest score. This method is easy to understand but unscientific; it may be useful primarily as an interim method.

Clearly, valuation of external environmental costs in the context of utility planning is an emerging field, one in which estimates and methods will evolve based on projects under way in various states, including New York, Massachusetts, and California. There is significant disagreement among experts on key methodological issues. Should costs be based on the payments that people are willing to make to avoid environmental damages or on the payments that must be made to them to accept these damages? What discount rates are appropriate? How should we estimate the effects and costs of different pollutants, especially those effects that are hard to price? (See Ottinger et al. 1990 for a full discussion of these issues.)

The extent of the disagreement about the magnitudes of environmental costs is reflected in Figure 1-2 (next page). The lower bars show the direct costs for a coal-fired baseload plant, gas-fired combined-cycle plant, and gas- and oil-fired combustion turbines. The low estimates of environmental costs are from the New York Public Service Commission, the high estimates from Chernick

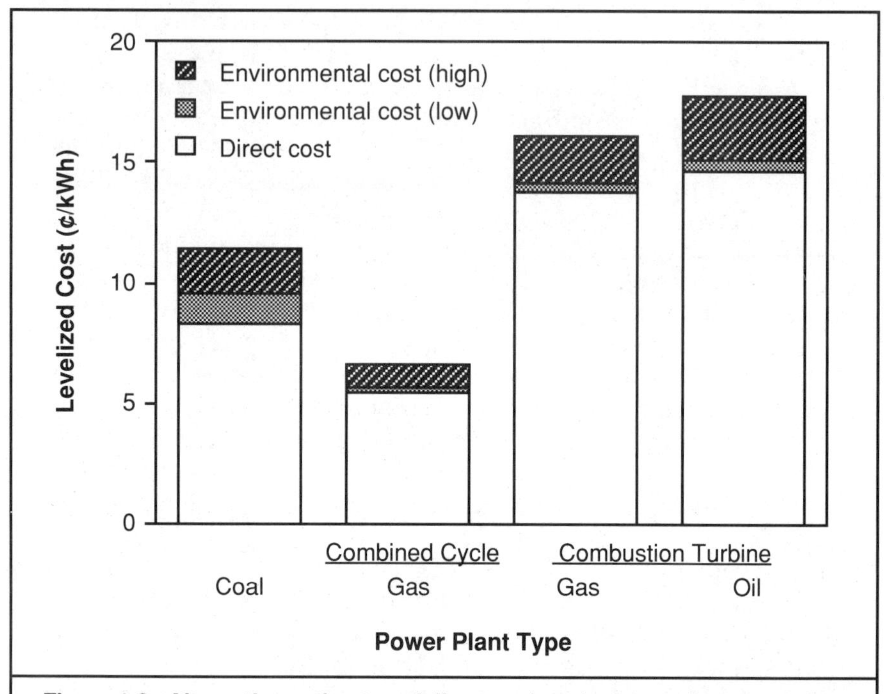

Figure 1-2. Alternative estimates of direct and environmental costs of electricity production from new power plants (Koomey 1990).

Note: The assumed capacity factor is 65% for the coal and combined-cycle plants and 10% for the combustion turbines.

(Koomey 1990). Interestingly, the low costs include the environmental impacts on water use and land use as well as air pollution, while the high estimates include air pollution only.

In the future, policies that consider environmental costs in resource acquisition will increasingly rely on site-specific methods whose results will depend on the particular location, technology, plant size, and fuel. This approach is in contrast to initial, more generic approaches. As a result, bidding processes will further differentiate among supply-side and DSM technologies based on individual project characteristics, including size, location, and expected emissions. New approaches, such as that adopted by the New York utilities in their bidding systems, may well be adopted by other states.

In addition, increased attention to environmental concerns may provide an important impetus for policy makers and PUCs to broaden the boundaries of IRP. PUCs may ask gas and electric utilities to compare the social costs and benefits of using gas directly with those of using gas-fired electricity for energy services such as water heating or cooking. Such analyses may significantly change the demands for electricity and gas. For example, both the policies of local air quality boards to limit vehicle emissions by encouraging electric cars and national legislation to reduce greenhouse gas emissions could affect future choices between electricity and gas.

Finally, electric utilities and others are likely to raise basic questions about the roles of PUCs and utilities in addressing environmental externalities versus the roles of federal and state agencies dealing with environmental quality (agencies such as the U.S. Environmental Protection Agency).

Bidding for Electric Power Resources

Nonutility power producers have become major sources of new generating capacity, principally because of the 1978 Public Utility Regulatory Policies Act (PURPA). Cogenerators and small power producers built nearly 15,000 MW of nonutility capacity during the 1980s. Under PURPA, PUCs are responsible for implementing pricing arrangements under which electricity is purchased from Qualifying Facilities (QFs) at the utility's avoided costs, which are determined administratively. To encourage QF suppliers, some states offered long-term contracts based on forecasts of avoided costs. In several states, the response by private producers was much greater than expected, partly because avoided-cost forecasts turned out to be high due to events in world oil and gas markets. Some utilities also claimed that the obligation-to-purchase provisions of PURPA and the open-ended nature of standard-offer contracts introduced substantial uncertainty about how much power would ultimately be developed. Thus, PURPA was not an unqualified success, because the supplier response created major planning and operational problems for some utilities.

During the past few years, some utilities and PUCs have experimented with competitive resource procurements (CRPs) as one way to obtain supply and DSM resources, partly in response to the problems associated with PURPA. Since 1986, about 25 solicitations have been issued by 14 utilities. Thus far, capacity offered by private producers has typically been 10–20 times greater than the utility's requirements. However, some utilities have received many bids that do not meet specifications and have therefore dropped these bids from serious consideration. For example, Central Maine Power received bids for over 2,300 MW of generating capacity in response to a 1989 solicitation; only about 1,000 MW remained as realistic options after the utility's initial review of the bids.

Private producers favor CRPs because the purchasing utility offers long-term contracts, which are needed to get financing on attractive terms. Competitive acquisition allows a utility to ration contracts for nonutility resources efficiently. Moreover, these contracts commonly transfer to private developers some of the risks associated with project siting and permitting, construction cost overruns, and environmental impacts. In addition, a competitive process can reduce the burden of estimating avoided cost by providing a market benchmark to determine value.

Despite these virtues, CRPs present formidable practical problems. Traditional utility planning requires trade-offs among financial, operating, and environmental features of resource alternatives. Competitive bidding requires the utility to address these issues through arms-length contracting. To assess bids, a utility must account for and assign values to multiple attributes of projects. This unbundling and explicit valuation of attributes is a new phenomenon in resource planning. Typically, utility bidding systems differentiate projects on pricing terms; operating characteristics; project status and viability (likelihood of successful development, for instance); and, in some cases, environmental impacts. Determining the economic value of these nonprice factors is probably the most difficult problem that utilities confront in designing bidding systems (Kahn et al. 1989).

In designing CRPs, two features are critical: (1) the method used to assess or score proposals, specifically the extent to which the utility discloses assessment criteria and the weight assigned to each feature before bid preparation, and (2) incorporation of DSM options into bidding schemes.

Bid Evaluation Criteria

Utilities have approached the bid solicitation and evaluation process in two ways. In the first approach, the utility's scoring system explicitly states the assessment criteria and weights for various features. Bidders score their projects themselves, assigning points in various categories, such as price, level of development, and dispatchability. PUCs and most utilities in Massachusetts, New Jersey, and New York rely on self-scoring systems.

Self-scoring is an open system in that the utility's bid evaluation process is transparent. Regulators can easily audit the utility's project rankings, minimizing controversy over the utility's selection of the winning bids. In addition, some PUCs favor self-scoring because it allows the regulators to shape utility planning decisions early in the resource-acquisition process rather than later in prudence reviews of contracts. However, some utilities are concerned that self-scoring denies them the flexibility needed to select the optimal mix of projects because self-scoring does not take into account a project's interactions with other projects. Considering projects independently is reasonable when the utility's resource need is small compared to existing capacity, but this practice becomes increasingly untenable when resource procurements are large relative to the existing utility system.

In the second approach, utilities do not state project selection criteria explicitly but describe them in qualitative terms only and provide only general guidance about preferences (Kahn et al. 1989). Bidders submit detailed proposals, which provide the basis for the utility's evaluation and ranking of projects. In this approach, the utility retains more discretion to select the optimal mix of projects and to negotiate with bidders in light of all offers received. In this closed approach, the utility has information about the evaluation process unavailable to bidders. Prominent examples of this approach include procurements issued by Virginia Power, Florida Power and Light, and Public Service of Indiana.

The closed approach acknowledges the complexity in optimizing resource selection when the values of proposed projects are multidimensional and uncertain, particularly when projected over long periods of time. This approach allows the utility to select the most efficient mix of bids, because it explicitly recognizes the interactive effects among individual projects and their effects on the utility system. Implicitly, PUCs that endorse this approach trust the utility's judgment. Utilities that want flexibility and discretion in bid evaluation and selection often exclude their subsidiaries from the bidding process, thus easing concerns about unfair competitive advantages.

Some utilities have combined elements of self-scoring and closed-bid systems. Central Maine Power includes elements of self-scoring but retains substantial flexibility to select attractive projects for further negotiation. Niagara Mohawk uses a self-scoring system for initial screening and then negotiates with bidders in the initial award group. The Massachusetts Department of Public Utilities recently proposed a similar approach.[5] These hybrid approaches

---

5. *Investigation by the Department on Its Own Motion into Proposed Rules to Implement Integrated Resource Management Practices for Electric Companies in the Commonwealth.* DPU 89-239, January 1990.

can successfully balance a utility's need for flexibility and discretion with the need to assure fairness.

Bid evaluation methods are an evolving art rather than a science. We expect continued experimentation with information requirements and risk sharing between utilities and private power producers.

Bidding for DSM Resources

CRPs raise still other issues, one of which is what types of bidders to include. For example, should QFs, independently owned generation facilities, energy service companies (ESCOs), large commercial and industrial customers, and the sponsoring utility itself all be allowed to bid?

Another issue is what types of resources to include. The appropriateness of including "saved kWh and kW" provided by ESCOs or individual customers has been vigorously debated (Cavanagh 1988; Cicchetti and Hogan 1989; and Joskow 1988 and 1990). Much of this debate has focused on problems of integrating DSM and supply resources in the same all-source bidding process and on the principles for determining the appropriate ceiling price for DSM resources (Goldman and Hirst 1989).

These debates raise interesting issues of how to measure the expected energy and demand savings, of whether all sectors and demand-side options should be included in a bidding program or whether the utility should target certain customer classes and end uses for a bidding program, and of how to integrate the DSM bidding program with a utility's own DSM programs. A key question is whether the ceiling price for DSM bids should be based on avoided cost or on the difference between avoided cost and average revenues (to reflect lost revenues). Some of these issues are not unique to DSM bidding and arise in utility DSM programs also.

Table 1-3 summarizes results from utilities that include DSM options in their bidding approach, including the megawatts offered by bidders and that selected by the utility. In addition, results are shown for recent supply-side procurements conducted by New England Electric System and Boston Edison, along with results from their DSM performance contracting programs involving ESCOs. Typically, 5–15 DSM bids have been submitted by ESCOs and individual customers. The DSM bidders have a stronger likelihood of winning (35–50%) than do supply-side projects. The amount of DSM savings proposed by winning bidders, while significant (10–47 MW over a three- to five-year period), represents a small part of a utility's overall DSM program for the same period (5–20%). Initial

Table 1-3. Supply and DSM resources in utility bidding programs (in MW).

Utility	Amount of Resource Requested	Supply Projects		DSM Projects	
		Proposed	Winning	Proposed	Winning
Central Maine Power	100	666	—	36	17
Central Maine Power	150–300	2,338	—	30	—
Orange & Rockland	100–150	1,395	141	29	18
Public Service Electric & Gas	200	654	210	47	47
Jersey Central	270	712	235	56	26
Puget Power	100	1,251	127	28	10
Separate auctions					
New England Electric	200	4,279	204	NA	14*
Boston Edison	200	2,800	200	NA	35*

*Savings from DSM performance contracting programs involving ESCOs

results reflect current infrastructure limitations in the ESCO industry, ESCOs' cautious responses to risks of guaranteeing savings, and ESCOs' limited experience with DSM bidding.

In summary, experience with incorporation of DSM options into bidding processes is limited. There are a few programs nationwide, although bidding programs are proliferating rapidly. Initial experience suggests that DSM bidding may have a limited role in a utility's overall DSM strategy but may not be appropriate for all market segments. For example, it is difficult to imagine DSM bids for the new construction market. The relative immaturity of the ESCO industry contrasts markedly with the strength of private power producers. In practice, this means that the quantities offered under DSM bidding programs will be small and will not reflect the full market potential of DSM. For utilities, DSM bidding programs may represent a potential business opportunity if utilities establish unregulated ESCOs. Several utilities have adopted this strategy, but most are skeptical about DSM bidding and prefer other ways to deliver DSM programs.

Developing Guidelines for Preparation of Utility Plans

A long-term resource plan is an opportunity for a utility to share its vision of the future with the public and to explain its plan to implement this vision. The quality of a utility resource plan should be judged from the results of implementing it. However, IRP is too new to have produced results, so the quality of plans based on IRP can be assessed only from what they say and how well they say it. This section therefore discusses guidelines for writing long-term resource plans, based on existing plans (Hirst et al. 1990; Schweitzer, Yourstone, and Hirst 1990). The purpose of these guidelines is to help both PUC staff who review utility plans and utility staff who prepare such plans. The plans prepared by Carolina Power and Light (1989), Green Mountain Power (1989), New England Electric System (1989), Northwest Power Planning Council (1989), Pacific Power & Light (1989), Puget Sound Power & Light (1989), and Seattle City Light (1989) contain many of the positive features in the guidelines.

The quality of a plan can be judged by at least four criteria:

- The clarity with which the contents of the plan, the procedures used to produce the plan, and the expected outcomes are presented
- The technical competence (including the computer models and supporting data and analysis) with which the plan was produced
- The adequacy and detail of the short-term action plan
- The extent to which the interests of various stakeholders are addressed

These criteria are explicated further in Table 1-4.

Clarity

The primary purpose of a utility's integrated resource plan is to help utility executives decide (and PUC commissioners review) which resources to acquire, in what amounts, and when. Thus, the report must be useful both within and outside the utility. The utility's plan should be well written and appropriately illustrated with tables and figures. The report should discuss the goals of the utility's planning process, explain the process used to produce the plan, present load forecasts (both peak and annual energy), compare existing resources with future loads to identify the need for additional resources, document the demand and supply resources considered, describe alternative resource portfolios, show the preferred long-term resource plan, and present the short-term actions to be taken in line with the long-term plan. Important decision points should be identified, and the use of monitoring procedures to provide input for those decisions should be explained. The most significant effects of choosing among the available options (capital and operating costs, resource availability, and environmental effects, for example)

Table 1-4. Checklist for a good integrated resource plan.

Clarity. Does the plan adequately inform various groups about future electricity resource needs, resource alternatives, and the utility's preferred strategy?

- Clear writing style
- Comprehensible to different groups
- Presentation of critical issues facing utility, its preferred plan, the basis for its selection, and key decisions to be made

Technical competence. Does the plan positively affect utility decisions on, and regulatory approval of, resource acquisitions?

- Comprehensive and multiple load forecasts
- Thorough consideration of demand-side options and programs
- Thorough consideration of supply options
- Consistent integration of demand and supply options
- Thoughtful uncertainty analyses
- Full explanation of preferred plan and its close competitors
- Use of appropriate time horizons

Action plan. Does the short-term action plan provide enough information to document utility's commitment to acquire resources in long-term plan, and to collect and analyze additional data to improve planning process?

Equity. Does the plan provide information so that different interests can assess the plan from their own perspectives?

- Adequate participation in plan development and review by various stakeholders
- Sufficient detail in report on effects of different plans

should be discussed. The report should also describe the data and analytical methods used to develop the plan. Finally, the plan should point the reader to more detailed documentation on these topics.

Technical Competence

In developing a plan, computer models are typically used for a variety of functions, such as load forecasting; screening, selection, and analysis of demand and supply resources; and calculations of production costs, revenue requirements, electricity prices, and financial parameters. These models are used to analyze a wide range of plausible futures and available resources in developing the utility's preferred resource portfolio. The basic structure of the models used, the assumptions upon which they are based, and the inputs utilized should be explained in the resource plans.

The technical competence of a utility's integrated resource plan is reflected most critically in the extent to which the demand and supply resources are presented as an integrated package. The analytical process used to integrate these different resources should be discussed. The criteria used to assess different combinations of resources—revenue requirements, annual capital costs, average prices, reserve margin, and emissions of pollutants—should be clearly stated.

Results for different combinations of supply and demand resources should be shown explicitly. It is not sufficient to treat demand as a subtraction from the load forecast and then do subsequent analysis with supply options only. Subtracting the effects of DSM programs from the forecast and using the resultant net forecast for resource planning eliminates DSM programs from all integrating analyses. This approach makes it difficult to assess the alternative combinations of DSM programs and supply resources and the uncertainties, risks, and risk-reduction benefits of DSM programs (small unit size and short lead time, for example). Demand-side resources must be treated in a fashion that is both substantively and analytically consistent with the treatment of supply resources; demand and supply resources must compete head to head.

The plan must show how the resource strategy will integrate key parts of the company: load forecasting, DSM resources, supply resources, finances, rates. It should also show the important feedback loops among these components (especially between rates and future loads).

The plan must analyze plausible future conditions, take into account uncertainties such as economic growth and fossil-fuel prices, and explain options for dealing with these conditions. It must also estimate the cost and performance of different resources, showing how utility resource-acquisition decisions are affected by these different assumptions and how different assumptions and decisions could affect customer and utility costs. Assumptions must be varied in consistent and plausible ways. Differences among resources in unit size, construction time, capital cost, and operating performance should be considered in terms of how they affect the uncertainties faced by utilities. Finally, the links between these multiple scenarios and the utility's resource-acquisition decisions must be demonstrated.

Action Plan

The action plan, in many ways the bottom line of the utility's plan, must be consistent with the long-term resource plan in order to ensure that the latter is implemented efficiently. The action plan also should be specific and detailed, identifying specific tasks for the next two to three years, along with organizational assignments, milestones, and budgets. The reader should be able to judge the utility's commitment to different actions from this short-term plan. For example, the action plan should show the number of expected participants and the expected reductions in annual energy use, summer peak, and winter peak for each DSM program. The action plan also should discuss the data and analysis activities—such as model development, data collection, and updated resource assessments—needed to prepare for the next integrated resource plan.

Equity

A final criterion is how actions recommended in the plan will affect interested parties. Stakeholders with different interests will be differently affected by short- and long-term costs, power availability, and other results of the utility's resource plan.

Without the involvement of customers and various interest groups, which requires two-way communication, a plan may ignore community needs. Accordingly, the plan should show that the utility sought ideas and advice from its customers and others in developing the plan. Energy experts from a state university, a state energy office, a PUC, environmental groups, and organizations representing industrial customers could be consulted as the plan is being developed. For example, utilities in New England are working closely with the Conservation Law Foundation to design, implement, and evaluate DSM programs (Ellis 1989).

Additional work is needed to refine the guidelines discussed here and to ensure that they are helpful to utilities and PUCs. In particular, PUCs should articulate better the reasons they want utilities to prepare such plans and how they, the PUCs, will use the plans in their deliberations. This articulation should avoid the data list or cookbook approach and focus on the purposes of the planning report. In the long run, the success of IRP should not be measured by assessing utility reports. Rather, the level and stability of energy-service costs, the degree of environmental protection, and the extent to which consensus is achieved on utility resource acquisitions will be important criteria.

Resource Planning at Gas Utilities

IRP is just beginning to be applied to the natural gas industry. At gas utilities, called local distribution companies (LDCs), resource planning addresses only options for purchasing and storing gas. Traditional planning for LDC resource acquisition seeks to purchase the lowest-cost and most reliable mix of supplies by determining the design day send out (the maximum amount of gas required for the coldest day in the coldest month), the provisions of various supply contracts, and available storage options.

LDC experience with DSM programs is limited to federally mandated programs, such as the Residential Conservation Service (created in 1978 and repealed in 1989) or low-income weatherization efforts mounted to create a positive image in the community or to reduce bill arrearages. Interruptible contracts with large industrial gas customers are also used by many LDCs to shave peaks (during very cold weather, for example).

The gas industry is concerned about declining sales and profits, largely because of experience during the1970s and 1980s, when gas customers adopted energy-conservation actions because of price increases and government programs. Gas consumption per customer fell 22% between 1974 and 1988 (EIA 1988). Many LDCs feel their focus now should be on increasing gas sales (rather than on encouraging conservation) because supplies of gas are ample and prices have been falling for several years.

Gas and Electric Industry Differences

Electric utilities are vertically integrated, while gas utilities are not. Production, transmission, and distribution of electricity are regulated primarily by PUCs, with the Federal Energy Regulatory Commission (FERC) involved only in wholesale contracts. The gas industry is organized and regulated differently. Natural gas is produced, transported, and distributed by three different sets of companies. Gas is produced by unregulated companies. Pipeline companies, regulated by FERC, move gas to local distribution companies, and LDCs, regulated by PUCs, distribute gas to consumers.

The time horizon for resource planning is generally closer in the gas industry than in the electric industry. Electric utilities, which construct power plants that last 30 to 40 years, plan accordingly. Gas utility planning depends on equipment lifetime, market conditions, and the length of contracts, which last less than 20 years.

Electric utilities meet load instantaneously and they ensure high reliability by maintaining extra generating capability rather than by storing electricity. LDCs, on the other hand, store gas and use interruptible contracts to maintain reliability for their core customers.

For electric utilities, procedures to determine avoided costs are reasonably well defined because of the decade of experience with PURPA. Avoided cost provides a benchmark against which to assess the value of resources offered by private producers and by DSM programs. Unlike electric utilities, gas LDCs have only a limited obligation to serve dual-fuel industrial customers, and this flexibility complicates the definition of avoided cost. The avoided-cost benchmark for natural gas, at least for the noncore market, is based on world oil prices. (See next paragraph for definition of "noncore market.") This benchmark is volatile and difficult to predict. Moreover, marginal-cost pricing is much less developed in the gas utility industry. For example, the marginal cost for gas could reflect limitations in pipeline capacity and alternative uses of the gas, say for generating electricity, both of which are upstream considerations for the LDC.

Gas Market Characteristics

The LDC market is divided into two segments, core and noncore. The core market consists of residential, commercial, and small industrial customers that depend entirely on the LDC for gas supplies. The noncore market consists primarily of large industrial customers and electric utilities, both of which can arrange for gas transportation and forgo purchasing gas from the LDC by purchasing gas directly from producers. In the noncore market, the LDC offers gas itself, in competition with gas marketers and independent producers, and local gas transportation, for which the LDC has a monopoly.

Recent reports on natural gas production capability show that today's abundant supply of gas may disappear within a year or two (American Gas Association 1990). As a consequence, LDCs will rely more on long-term contracts and less on spot-market purchases for their gas supplies. Significant regional differences in gas supply and prices are likely to persist, however, because of differences in pipeline capacity and distance between gas supplies and markets.

Regulatory Environment

Throughout the 1980s, the gas industry and PUCs had little interest in applying IRP because resource planning for electric utilities dominated PUC interest, regulators emphasized deregulation of the gas industry, and estimates of gas resources seemed to assure an adequate supply of gas at low cost, reducing the importance of long-term planning.

However, regulators in a few states—Nevada, Washington, and Wisconsin—and in the District of Columbia are beginning to examine

IRP for the gas industry. PUCs are interested in gas planning for the following reasons:

- Benefits of electric-utility IRP, especially in implementing DSM programs
- Recent requests for investments in new pipeline
- Possible environmental benefits of using gas instead of electricity
- Interest in fuel switching, including the use of gas air-conditioning technologies to cut electric system peaks in summer

Because IRP is often viewed skeptically by gas utilities, efforts to date have been started by regulators. Gas utility experience with IRP is often limited to residential retrofit programs. These programs are generally not based on reliable forecasts of future gas demands, on testing (via pilot programs) of DSM marketing and incentive mechanisms, or on evaluation of existing DSM programs.

To achieve the next level of program development, more rigorous analysis should be conducted to quantify the DSM potentials in specific market segments. These analyses might employ end-use models to forecast gas demands for each customer class.To date, few end-use data are available except in the areas of residential retrofits (Ternes et al. 1990) and some new technology applications (Brodrick and Patel 1990; GRI 1989). DSM is important for commercial and industrial customers with interruptible service or dual-fuel capability (or both) when gas supply is limited or gas costs become too high because of extreme weather conditions.

Almost all current analyses examine least-cost purchasing, selecting the best mix of supply and storage options to achieve low prices for consumers and high earnings for shareholders (McDermott 1987). Only a few methods—one being the California Standard Practice Manual (California Commissions 1987)—assess the cost-effectiveness of gas utility DSM programs.

The next stage following analysis would be the integration of demand and supply options to assess the best resource mix.

To advance gas IRP, several key questions need to be addressed. What effect will DSM programs have on LDC supply reliability and profitability? What are the economic implications of electric-utility cost-effectiveness tests—societal, all ratepayer, participant, and no-losers—to LDCs, their customers, and shareholders? How should fuel switching be included in gas IRP? Can gas-utility DSM programs be used to reduce industrial bypass? What regulatory adjustments are necessary to encourage gas IRP while maintaining company profitability?

Developing Federal Policies to Promote IRP and DSM

Because electricity production consumes almost 40% of the primary energy used in the U.S., electricity must be a major part of national energy policy. In addition, concerns about environmental quality, economic productivity, international competitiveness, and national security suggest a larger role for the federal government in working with utilities to expand their planning and to implement DSM programs.

Improving energy efficiency through utilities may be a particularly effective way to reach millions of U.S. energy consumers. Utilities have direct monthly contact with all their customers through meter reading and billing and are usually well-respected organizations in their communities. Thus, the federal government can work with a few hundred utilities and, through them, reach tens of millions of households and millions of businesses.

Recommendation #1: Require FERC to Incorporate DSM Programs in Utility Wholesale Contracts. FERC approves all wholesale transactions among utilities. Currently, FERC reviews of proposed contracts do not consider energy efficiency. Prior to approving a contract, FERC could require the buying utility to show that it has acquired all the conservation and load-management resources in its service area that cost less than the proposed purchased power. An integrated resource plan approved by the utility's PUC could satisfy this requirement, and a state-approved plan would eliminate concerns that FERC was preempting state regulation. Implementing such an expanded review of wholesale contracts might require modification of the Federal Power Act.

Recommendation #2: Require Federal Electric Utilities to Expand DSM Programs. DOE's Power Marketing Agencies (PMAs) and Tennessee Valley Authority (TVA), an independent federal corporation, account for one-tenth of the electricity consumed in the U.S. Traditionally, TVA and the Bonneville Power Administration (BPA), the largest PMA, have operated large DSM programs, which saved energy for their customers and served as examples for other utilities. Unfortunately, short-term budget considerations forced reductions in these programs at both agencies. Indeed, TVA cancelled all of its conservation programs in 1989. BPA, on the other hand, plans to increase its conservation budgets over the next several years.

New legislation could require these federal power authorities to expand their DSM programs and to explicitly consider environmental and social factors in their benefit/cost analyses of all resource alternatives. Such legislation would be a logical extension of the 1980 Pacific Northwest Electric Power Planning and Conservation Act (P.L. 96-501), which explicitly made conservation the electricity resource of choice and gave it a 10% bonus to be used in economic analyses of alternative resources. The 10% bonus reflects the environmental and social benefits of conservation compared to supply resources. The other federal utilities could employ similar factors in their resource assessments.

Recommendation #3: Expand DOE Technology-Transfer Activities to Utilities. DOE's Integrated Resource Planning Program manages projects aimed at improving the long-term resource-planning process and tools (data and analytical methods) used by utilities (Goldman, Hirst, and Krause 1989; Berry and Hirst 1989). DOE sponsored conferences on utility planning in 1988, 1989, 1990, and 1991. The DOE program could be expanded to fund additional cooperative projects with utilities and PUCs. This approach focuses on cost-sharing projects, with DOE assistance provided through the DOE national laboratories and other government contractors.

Many innovative and successful IRP programs operate throughout the country. However, information on these successes is hard to get because sponsoring utilities and PUCs have little incentive to publish it. Thus, DOE can play a valuable role by participating in these programs, ensuring that they are carefully evaluated, and then funding preparation of reports and conference presentations that effectively disseminate information to other utilities and state agencies. The Northeast Region Demand-Side Management Data Exchange (NORDAX), funded in part by DOE, is a good example of such technology transfer. The initial phase of NORDAX, a consortium of more than 20 utilities, yielded a database with information on 90 DSM programs operated by 17 utilities in the region (Camera, Stormont, and Sabo 1989). Another critical area for DOE attention is the transfer of planning methods, data, and processes from electric to gas utilities.

Recommendation #4: Collect More Information on Energy Use. The Energy Information Administration (EIA) is responsible for collecting, evaluating, analyzing, and disseminating information on energy reserves, production, demand, and technologies. EIA focuses on the supply of, rather than the demand for, energy. For example, EIA's *Annual Energy Review* (EIA 1989b) contains separate chapters on fossil-fuel reserves, petroleum, natural gas, coal, electricity, nuclear energy, renewable energy, and financial indicators, but only one chapter on the entire subject of energy consumption.

EIA (1989a) collects detailed information from electric utilities on individual power plants: construction costs and capacities; annual operations and maintenance expenses; and monthly fuel consumption, generation, availability, and emissions. Data collected by FERC—through FERC Form 1, for example—are similarly detailed with respect to electricity production and disposition; purchases and sales; construction costs and operations for power plants; and costs and characteristics of transmission lines, substations, and transformers. Unfortunately, EIA and FERC collect no comparable data on utility energy-efficiency and load-management programs. EIA and FERC could expand the data-collection forms completed by utilities (both electric and gas) to require information on utility DSM programs. This addition would help to redress the imbalance between the supply and demand sides in data-collection activities.

Conclusions

More than half of the primary energy consumed in the U.S. flows through electric and gas utilities. Therefore, the economic and environmental effects of utility actions are enormous. IRP represents a new way for utilities to meet the energy-service needs of their customers. Because IRP is a comprehensive and open process, its implementation is likely to yield large benefits in terms of an optimized mix of resources and fewer controversies over utility decisions.

Much work is needed to convert the potential benefits of IRP into reality. This paper dealt with a few of the most important topics, including changes in regulation to align the interests of utility customers with those of utility shareholders, incorporation of environmental factors into resource planning, bidding for resources, guidelines for review of utility plans, planning for gas utilities, and increased activities by the federal government.

References

American Gas Association. 1990. *Natural Gas Production Capability—1990*. Arlington, Va.: American Gas Association.

Argonne National Laboratory. 1989. *Proceedings for the 1989 Conference on Energy Conservation Program Evaluation: Conservation and Resource Management*. Argonne, Ill.: Argonne National Laboratory.

Berry, L., and E. Hirst. 1989. *Recent Accomplishments of the U.S. Department of Energy's Least-Cost Utility Planning Program*. ORNL/CON-288. Oak Ridge, Tenn.: Oak Ridge National Laboratory.

Brodrick, J., and R. Patel. 1990. "Defining the Market for Gas Cooling." *ASHRAE Journal* 32 (January): 24–31.

California Public Utilities Commission and California Energy Commission. 1987. *Standard Practice Manual, Economic Analysis of Demand-Side Programs*. P400-87-006. San Francisco: California Public Utilities Commission.

Camera, R., D. Stormont, and C. Sabo. 1989. "Developing Reliable Data on DSM Programs: The NORDAX Experience." Chap. 37 in *Demand-side Management Strategies for the 90s, Proceedings of the Fourth National Conference on Utility DSM Programs*. CU-6367. Palo Alto, Calif.: Electric Power Research Institute.

Carolina Power and Light. 1989. North Carolina Utilities Commission, Docket No. E-100, Sub 58 (April). CP&L Exhibits 1, 2, and 3. Raleigh, N.C.

Cavanagh, R. 1986. "Least-Cost Planning Imperatives for Electric Utilities and Their Regulators." *Harvard Environmental Law Review* 10 (2): 299–344.

——. 1988. "The Role of Conservation Resources in Competitive Bidding Systems for Electricity Supply." Testimony before the Subcommittee on Energy and Power, U.S. House of Representatives, on 31 March, Washington, D.C.

CEC. See California Public Utilities Commission and California Energy Commission.

Chernick, P., and E. Caverhill. 1989. *The Valuation of Externalities from Energy Production, Delivery, and Use*. Boston: Boston Gas Company.

Cicchetti, C., and W. Hogan. 1989. "Including Unbundled Demand-side Options in Electric Utility Bidding Programs." *Public Utilities Fortnightly* 123 (8 June): 9–20.

Cohen, S., J. Eto, C. Goldman, J. Beldock, and G. Crandall. 1990. *A Survey of State PUC Activities to Incorporate Environmental Externalities into Electric Utility Planning and Regulation*. LBL-28616. Berkeley, Calif.: Lawrence Berkeley Laboratory.

CP&L. See Carolina Power and Light.

CPUC. See California Public Utilities Commission.

Destribats, A., J. Lowell, and D. White. 1990. "Dispatches from the Front: New Concepts in Integrated Planning." Westborough, Mass.: New England Power Service Company.

EIA. See Energy Information Administration.

Electric Power Research Institute. 1988. "Status of Least-Cost Planning in the United States." EM-6133. Palo Alto, Calif.: Electric Power Research Institute.

Ellis, W. 1989. "The Collaborative Process in Utility Resource Planning." *Public Utilities Fortnightly* 123 (22 June): 9–12.

Energy Information Administration. 1988. *Natural Gas Annual*. DOE/EIA-0131 (88): 2. Washington, D.C.: U.S. Department of Energy.

——. 1989a. *Directory of Energy Data Collection Forms*. DOE/EIA-0249 (88). Washington, D.C.: U.S. Department of Energy.

——. 1989b. *Annual Energy Review 1988.* DOE/EIA-0384 (88). Washington, D.C.: U.S. Department of Energy.

EPRI. See Electric Power Research Institute.

Gas Research Institute. 1989. *Assessment of Large Tonnage Gas-Fired Cooling Technologies for the Commercial Sector.* Report 88-D162. Chicago: Gas Research Institute.

Gellings, C., J. Chamberlin, and J. Clinton. 1987. *Moving Toward Integrated Resource Planning: Understanding the Theory and Practice of Least-Cost Planning and Demand-Side Management.* EM-5065. Palo Alto, Calif.: Electric Power Research Institute.

Goldman, C., and E. Hirst. 1989. "Key Issues in Developing Demand-Side Bidding Programs." Chap. 23 in *Demand-Side Management: Partnerships in Planning for the Next Decade, Proceedings of ECNE National Conference on Utility DSM Programs.* CU-6598. Palo Alto, Calif.: Electric Power Research Institute.

Goldman, C., E. Hirst, and F. Krause. 1989. *Least-Cost Planning in the Utility Sector: Progress and Challenges.* LBL-27130. Berkeley, Calif.: Lawrence Berkeley Laboratory. (Or ORNL/CON-284. Oak Ridge, Tenn.: Oak Ridge National Laboratory.)

Green Mountain Power. 1989. *Integrated Resource Plan.* South Burlington, Vt.: Green Mountain Power Corp.

GRI. See Gas Research Institute.

Hirst, E. 1988a. "Meeting Future Electricity Needs." *Forum for Applied Research and Public Policy* 3 (fall).

——. 1988b. "Integrated Resource Planning: The Role of Regulatory Commissions." *Public Utilities Fortnightly* 122 (15 September): 34–42.

Hirst, E., and M. Schweitzer. 1990. "Electric-Utility Resource Planning and Decision-Making: The Importance of Uncertainty." *Risk Analysis* 10 (1): 137–146.

Hirst, E., M. Schweitzer, E. Yourstone, and J. Eto. 1990. *Assessing Integrated Resource Plans Prepared by Electric Utilities.* ORNL/CON-298. Oak Ridge, Tenn.: Oak Ridge National Laboratory.

Hobbs, B., and P. Maheshwari. 1990. "A Decision Analysis of the Effect of Uncertainty Upon Electric Utility Planning." *Energy* 15 (September): 785–802.

Joskow, P. 1988. Testimony before the Subcommittee on Energy and Power, U.S. House of Representatives, 31 March, Washington, D.C.

——. 1990. "Understanding the 'Unbundled' Utility Conservation Bidding Proposal." *Public Utilities Fortnightly* 125 (4 January): 18–28.

Kahn, E., C. Goldman, S. Stoft, and D. Berman. 1989. *Evaluation Methods in Competitive Bidding for Electric Power.* LBL-26924. Berkeley, Calif.: Lawrence Berkeley Laboratory.

Koomey, J. 1990. *Comparative Analysis of Monetary Estimates of External Environmental Costs Associated with Combustion of Fossil Fuels.* LBL-28313. Berkeley, Calif.: Lawrence Berkeley Laboratory.

Krause, F., and J. Eto. 1988. *The Demand Side: Conceptual and Method-*

ological Issues. Vol. 2 of *Least-Cost Utility Planning, a Handbook for Public Utility Commissioners*. Berkeley, Calif.: Lawrence Berkeley Laboratory.

McDermott, K. 1987. *Wrestling with Transition: Least-Cost Planning in the Natural Gas Industry*. Chicago: Center for Regulatory Studies.

Marnay, C., and G. Comnes. 1990. *Ratemaking for Conservation: The California ERAM Experience*. LBL-28019. Berkeley, Calif.: Lawrence Berkeley Laboratory.

Moskovitz, D. 1989. "Profits & Progress through Least-Cost Planning." Washington, D.C.: National Association of Regulatory Utility Commissioners.

NARUC. See National Association of Regulatory Utility Commissioners.

National Association of Regulatory Utility Commissioners. 1988. *Least-Cost Utility Planning, a Handbook for Public Utility Commissioners*. Washington, D.C.: National Association of Regulatory Utility Commissioners.

——. 1989. "Resolution in Support of Incentives for Electric Utility Least-Cost Planning." Washington, D.C.: National Association of Regulatory Utility Commissioners.

NEES. See New England Electric System.

New England Electric System. 1989. *Conservation and Load Management Annual Report*. Westborough, Mass.: New England Electric System.

Northwest Power Planning Council. 1989. *1989 Supplement to the 1986 Northwest Conservation and Electric Power Plan*. Vol. 2. Portland, Ore.: Northwest Power Planning Council.

NPPC. See Northwest Power Planning Council.

Ontario Hydro. 1990. *Providing the Balance of Power: Environmental Analysis*." Toronto: Ontario Hydro.

Ottinger, R., D. Wooley, N. Robinson, D. Hodas, and S. Babb. 1990. *Environmental Costs of Electricity*. Dobbs Ferry, N.Y.: Oceana Publications.

Pacific Power & Light and Utah Power & Light. 1989. *Planning for Stable Growth*. Portland, Ore.: Pacific Power & Light.

Puget Sound Power & Light. 1989. *Demand and Resource Evaluation: Securing Future Opportunities, 1990–1991*. Bellevue, Wash.: Puget Sound Power & Light.

Schultz, D. 1990. "Utility Shareholder Incentive Mechanisms: A Technical Review and Comparison of Shared Savings Proposals for Demand-Side Management Programs." Staff report, Division of Ratepayer Advocates, California Public Utilities Commission, Sacramento, Calif.

Schweitzer, M., E. Yourstone, and E. Hirst. 1990. *Key Issues in Electric Utility Integrated Resource Planning: Findings from a Nationwide Study*. ORNL/CON-300. Oak Ridge, Tenn.: Oak Ridge National Laboratory.

Seattle City Light. 1989. "Strategic Corporate Plan 1990–1991." Seattle:

Sergel, R. 1989. Testimony and exhibits submitted on behalf of Granite State Electric Company. Westborough, Mass.: New England Power Service Company.

Ternes, M., P. Hu, L. Williams, and P. Goewey. 1990. *The National Fuel End-Use Efficiency Field Test: Energy Savings and Performance of an Improved Energy Efficiency Measure Selection Technique.* ORNL/CON-303. Oak Ridge, Tenn.: Oak Ridge National Laboratory.

Wiel, S. 1989. "Making Electric Efficiency Profitable." *Public Utilities Fortnightly* 124 (6 July): 9–16.

Chapter 2

Consideration of Environmental Externality Costs in Electric Utility Resource Selections and Regulation

Richard L. Ottinger, *Pace University Law School Center for Environmental Legal Studies*

Approximately half the state electric utility regulatory commissions in the United States have started to require utilities to consider environmental externality costs in planning and resource selection. The principal rationale for doing so is that electric utility operations impose very real and substantial damage to human health and the environment, damage not taken into account either by traditional utility least-cost planning and resource selection procedures or by government pollution regulations. It is becoming clear that, given the high likelihood of more stringent controls in the future, failing to consider environmental externality costs when selecting resources is imprudent.

Most regulatory commissions requiring utilities to consider environmental externalities have left it to the utilities to compute the societal costs of residual environmental effects. Some commissions, however, have either set those costs themselves or used a proxy adder for the costs of polluting resources or a bonus for nonpolluting resources. These commissions have used control or pollution mitigation costs, rather than societal damage costs, in their regulatory computations.

This chapter recommends using damage costs where adequate studies exist to permit quantification; it also discusses methodologies for measuring damage costs and describes the means for incorporating such costs into the resource selection process.

Background

Internalizing the environmental costs imposed on society by polluters is the wave of the future in addressing environmental pollution. Governments are just starting to consider supplementing pollution regulations with pollution taxes or fees that would add to market prices the social costs of damages inflicted by polluting resources. The Organization for Economic Cooperation and Development (OECD) has recently published a review of pollution levies, indicating a total of 85 pollution taxes in six of its principal countries (Economist 1989). Germany, France, and Holland impose wastewater effluent charges ($2, $9, and $39 per capita respectively); Switzerland imposes extra landing fees on noisy aircraft; and Sweden and Norway require returnable deposits on automobile bodies to prevent their being dumped. West Germany is considering an auto tax based on tail pipe exhausts (ibid.). The U.S. House of Representatives Ways and Means Committee has held hearings on pollution taxes (6, 7, 14 March 1990).

Government regulation of pollution has not adequately addressed the severe threats to the planet posed by global warming, acid rain, urban smog, and other forms of toxic contamination of our air, water, and food supplies. These environmental insults have taken place despite the array of regulations designed to control them. Economic growth, in both developing and industrialized countries, has received higher government priority than pollution control and has outpaced governmental efforts to require pollution control technology or a switch to less polluting industrial resources. For example, in the United States, even after implementation of the Clean Air Act revisions recently passed by Congress (with all their attendant political compromises), substantial environmental costs still will be imposed on society by residual, uncontrolled impacts, many of which would be economic to mitigate.

The marketplace could powerfully influence the environmental effects of industrial decision making. If industry were required to pay the costs of the pollution it imposes on society, economics could induce industry to choose more environmentally benign resources.

Environmental organizations, which historically have resisted pricing environmental impacts on grounds that so doing would constitute a license to pollute, have now embraced the idea. The Environmental Defense Fund's Daniel Dudek, a leading advocate of letting the market reflect the costs of pollution, helped draft the Administration's Clean Air Act proposal to create emissions trading rights.

Traditionally conservative electric utilities and state utility regulatory commissions have pioneered in applying marketplace principles to valuation of environmental externalities. In 26 jurisdictions, utility regulatory commissions have started to consider whether and how to incorporate these externalities into utility planning and resource selection procedures.

Until recently, most utilities, with the approval of state regulatory commissions, have selected supply- and demand-side resources on a least-cost basis, without considering the environmental costs that persist even after utilities have complied with all applicable environmental regulations. This practice effectively assigns a value of zero to residual environmental externality costs, which may be substantial.

As difficult as it is to compute the exact costs of environmental damages, "a crude approximation, made as exact as possible and changed over time to reflect new information, would be preferable to the manifestly unjust approximation caused by ignoring these costs" (Bland 1986).

Why Utilities and Commissions Should Consider Environmental Externality Costs

Incorporating the monetary costs of environmental externalities into utility resource selection procedures will encourage utilities to invest in more environmentally benign resources. Nevertheless, it can be argued that environmental externalities should be addressed solely by pollution controls or by legislated taxes or fees rather than by utility actions and regulatory commission orders affecting resource selection. Indeed, environmental costs of other industries, such as automobile and chemical manufacturing and smelting, are addressed only through legislated regulation. National pollution fees could internalize environmental externality costs for all polluting sources, and thereby send correct price signals to the marketplace. Likewise, regulations eliminating all economically unacceptable pollution would also internalize environmental externality costs.

Neither action is likely, however. Complete economic pollution controls are likely to be politically infeasible; national pollution taxes or fees are unlikely because the administration and Congress have been reluctant to impose new taxes. Furthermore, the command-

and-control structure of existing environmental regulations ignores cost-effective pollution prevention opportunities. National policy makers committed to this form of regulation would, in the presence of political forces, select a suboptimal level of environmental quality.

Assuming that legislation does not compel utilities to internalize environmental externality costs, utilities will go on selecting resources that, after regulation, still impose significant environmental damages on society. This practice will continue unless environmental costs are included in these considerations, because cost is the principal criterion for selecting a resource.

There are two justifications for asking utilities to factor externalities into their resource selection processes: (1) utilities are franchised monopolies vested with a public interest that includes environmental protection;[1] (2) foreseeable international, federal, and state environmental laws and regulations are likely to impose more stringent environmental controls over the 30- to 40-year life span of electric power plants, making it imprudent for utilities to invest in resources that will have to be abandoned or retrofitted at great expense. The traditional role of utility regulatory commissions includes overseeing utility public interest obligations and preventing imprudent investments.

Note that when commissions require consideration of environmental externalities in utility planning and resource selection, they are not internalizing these costs but merely ensuring prudent selection of new resources. This policy is a good prelude to actual internalization.

Costs to Be Included

Commissions and utilities that decide to consider externalities face the daunting task of determining those costs. The first question they must answer is what kinds of costs to include. Most commissions that have addressed externalities have determined that only environmental damages should be included. However, the most thorough study of externality costs, completed recently for the European Economic Community, seeks to value not only environmental damage but also impacts on production, employment, and trade balance;

---

1. The state of New York, for example, requires all state agencies to minimize or avoid environmental impacts "to the maximum extent practicable." See, for example, the New York Environmental Quality Review Act (SEQRA), New York Environmental Conservation Law, Section 8-0107, and its implementing regulation, 6 NYCRR, Part 617.9 (c) (3).

depletion of nonrenewable resources; public subsidies and R&D expenditures; and "induced public expenditures" such as defense costs (Hohmeyer 1989).

The most severe environmental costs imposed on society by electric utility operations derive from the risk of damages to human health and the environment from air pollutants emitted by fossil-fuel–fired generation and from the radiation emitted by nuclear plant operations.

The principal culprits among air pollutant emissions are the greenhouse gases, most significantly carbon dioxide (CO_2); sulphur dioxide (SO_2) and nitrogen oxides (NO_x), the principal precursors of acid rain; tropospheric ozone resulting from chemical interactions of NO_x and volatile organic compounds in the presence of sunlight; and particulates, which provide the medium for ingestion and inspiration of toxic co-pollutants. Nationally, electric utilities accounted in 1988 for 33% of CO_2 emissions (Machado and Piltz 1988) and in 1985 for 68% of SO_2 emissions (16.2 out of 23.7 tons) and 33% of NO_x emissions (7.0 out of 21.1 tons) (NAPAP 1987).

Many of the above air pollutants react synergistically after release, forming new chemical combinations that react together to inflict increased environmental damage. Furthermore, emission damages from future power plants will be cumulative, not simply additive, to the damages from pollutants already emitted by existing plants: that is, they will do more damage than they would have done if no previous pollution had been present. These synergistic and cumulative effects, not just the additive damages from each new pollutant, should be considered.

Risks of nuclear-powered resources include radiation from uranium milling and mining, low-level radiation in and around operating plants and accidental contamination of plant personnel, catastrophic accidents such as occurred at Chernobyl and Three Mile Island, contamination from plant decommissioning, disposal of mill tailings and of high- and low-level nuclear waste, and impact on fish at nuclear facilities. These risks are largely unaccounted for under current regulation. A recent British report asserts that the nuclear cycle also emits carbon dioxide at levels potentially comparable to fossil-fuel plants (Hill 1990).

In addition to damage from air pollution and risks from radiation, societal costs imposed by many electric service supply resources may include water pollution, land deprivation, agricultural losses (from flooding by dams, for example), and contamination from solid waste disposal. Damage from electromagnetic fields has also been asserted. These costs are generally much less than those resulting from air pollution and radiation damage, but they are significant enough to merit consideration.

Air- and water-polluting emissions from generating plants often extend beyond the state or country in which the plant is located. Most state commissions that have addressed environmental impact costs have included all costs to society, not just those affecting their own states.[2]

Environmental externality costs should also reflect differences between pollution costs in urban and rural settings; where appropriate, emissions should be calculated per unit of population. Pollutants emitted near a heavily populated metropolis will produce vastly greater human health costs than those emitted near unpopulated areas. Similarly, agricultural damage costs will be higher where emissions are deposited on farming communities rather than on urban areas.

Costs from the entire fuel cycle should be considered (front-end, operational, and back-end costs). It is difficult to know how far to pursue front-end costs, of course. One could go back infinitely far, estimating the social costs of manufacturing all the equipment and machinery necessary to manufacture the equipment and machinery, and so on, for each stage of the fabrication process. At the least, the first-generation costs of plant construction and fuel transport and the production costs of demand-saving or renewable equipment and facilities should be considered.

Control versus Damage Costs

Having decided which costs to include, the next major problem facing a commission or utility is how to calculate them. The first major issue is whether to use the costs of actual damages or the costs of controlling pollutants before the damage occurs. There is considerable difference among experts on this subject.[3]

The advantage of using control costs is that readily available data make these costs easier to determine and thus more defensible. Costs derived from legislative standards, it can be argued, reflect the level of protection the relevant agency experts and the legislative body consider safe, though in fact standards tend to be

2. Vermont is the only exception, accounting for only that portion of out-of-state pollution from Vermont power plants that impacts Vermont residents. However, the Vermont commission seems to be backing off this position in its most recent proceedings: Vermont PSB, *Re Least-Cost Investments, Energy Efficiency, Conservation, and Management of Demand for Energy,* Docket No. 5330.
3. For arguments supporting use of control costs, see Chernick and Caverhill 1989; for the case for using damage costs, see EPRI 1988.

determined politically rather than scientifically. The relative ease of determining control costs has caused all 26 state commissions that have ordered consideration of environmental externality costs to date to use them both as the basis for quantification of such costs and, where used, in the calculation of adders to the cost of polluting resources.[4]

The disadvantage of using control costs is that they may bear little or no relationship to the actual costs of damages imposed on society by the relevant pollutants, and they seldom cover all the risks involved because elected representatives tend, for political reasons, to enact controls well below marginal damage costs. Furthermore, control standards such as those established by the National Ambient Air Quality Standards of the Clean Air Act are adopted at a level to protect the public health and welfare with "an adequate margin of safety," often without regard to the costs to society of health and welfare damages. On the other hand, control costs might exceed damage costs, as in cases where standards are set to protect the most sensitive individuals in society. And there are many power plant pollutants, such as CO_2, for which no standards have been set.

The main advantage of using damage costs is that they are the relevant costs to be considered. It is the risk of damage to society, rather than the cost of controls, that the practice of incorporating environmental externality costs into utility resource selection seeks to address. Damage costs are useful as well for determining how much it is worth spending to institute additional controls.

The main disadvantage of using damage costs is the difficulty of calculating and defending them. Some experts feel that utility regulatory commissions (as opposed to agencies with environmental expertise like EPA) would have difficulty dealing with technical matters like valuation of human life and nonmonetarized costs like valuation of recreational facilities. On the other hand, there are some adequate scientific studies. Defense of a legal challenge to these values should be no more difficult than the generally successful defense of EPA health and safety standards that are based on similar kinds of scientific studies.

In view of the advantages and disadvantages described above, the most beneficial policy is to use damage costs when feasible,

4. The Oregon Public Service Commission, however, has ordered its utilities to seek to quantify damage costs in evaluating resource selections, *Re Least Cost Planning*, UM 180, OR PSC Order No. 89-507 (20 April 1989). The New York Public Service Commission has ordered a pooled study by its utilities of their environmental externality damage costs, NY PSC Case 28223, *Electric Utility Conservation Programs*, Opinion and Order 89-15 (23 May 1989).

since these costs are most relevant to the impacts on society. Where adequate studies exist, damage costs should be used. Where studies on damage costs are inadequate, as in the case of global warming research, then control costs should be used as the best available substitute, far superior to setting damage costs at zero by ignoring them.

Where impending legislative controls are reasonably ascertainable, as was the case with the Clean Air Act amendments, the effects of the new controls on damage costs must be taken into account, since the pollutant costs covered will be internalized once the legislation is enacted. Of course, if controls like scrubbers or bag houses (a technology for removing particulates from flue gas) are required by statutory or regulatory mandate, the costs imposed on society by the pollutants controlled will no longer be external costs.

Major Issues in Damage Risk Valuation

General Considerations

In the valuing of environmental damages, the objective is to determine the risk of damage rather than to assess the damages themselves. This is a most important principle. We are seeking to define the cost of the *risk* to life, health, and the environment, and the costs that people are willing to pay to avoid such risks or assume them.

For example, it would be inappropriate to measure mortality damages by seeking to measure the value of a human life, say by adding up the reasonably expected lost lifetime earnings of the individual or individuals affected. The value would vary by earning power, with rich people valued more than poor ones and with housewives and the elderly considered to be of negligible value. Similarly, it is inappropriate to use only mitigation costs as a measure of externality values. Adding up the doctors' bills is inadequate, for example, in valuing human health damages. Who among us would be willing to incur the hardships associated with cancer or debilitating injury even if fully compensated for the cost of treatment? But for each individual or population of individuals, it would be appropriate to measure the value of the risk posed to their lives by determining what they would be willing to pay to avoid the risk or what they would be willing to be compensated to assume the risk.

Also, risks must be quantified for populations rather than for individual people, crops, or animals. An event that kills or harms a very small number of individuals may be very costly to them, but the social costs of this damage will be insignificant. This distinction does not have anything to do with the potential criminal or civil liability for taking an individual life or with the value society places on every human being; it has only to do with valuation of risks to life for purposes of influencing utility resource selections. The value of the risk of loss to a very few individuals simply is not large enough to affect the economics of choosing one kind of utility resource over another.

Likewise, damage awards by a judge or jury for particular environmental damages are generally not very useful because they value the harm to an individual of an actual event, not the risks to populations of possible events. Also, particularly in the case of jury awards, damage awards are not scientifically derived.

Measurement Methodologies

Assigning monetary values to risks poses the following issues:

- Which measurement methodologies to use (how and when to use market prices, revealed preference, hedonic pricing, awards, or contingent valuation)
- How to allocate costs to joint projects
- What discount rates and real value escalation to apply
- How to adjust for uncertainty

(For a good discussion of all the externality costing methodologies and their applications, advantages, and disadvantages, see EPRI 1988; Freeman 1979; and, with respect to contingent valuation, Mitchell and Carson 1989.)

Market prices, where available, are useful in determining environmental damages. Knowing there is a 100% risk that a particular crop will be affected by a power plant and knowing the extent of the harm that will be imposed, one can multiply the crop loss by the market value to obtain the damages. Unfortunately, in most cases, risk and yield loss are seldom known with certainty, and, with large losses, the market price may be affected by the loss.

Revealed preference values are based on observed behavior. They are derived from the costs individuals, by their actions, have revealed they are willing to pay to avoid environmental damages or to accept as compensation for suffering such damages. Thus, in the case of loss of fishing opportunities in a lake because of acid deposition, travel costs fishermen are actually paying to reach alternative fishing areas might be used to value the damages. Travel costs, however, fail to take into account values not encompassed by the particular behavior measured. For example, travel costs would not accurately value the destruction of a unique historic resource even though there were other historic resources that could be visited. They may also fail to account for characteristics, such as an individual's age or income, that might prevent his or her traveling to alternative sites.

Hedonic pricing is a form of revealed preference that uses market-based prices to infer prices of nonpriced goods and services. For example, selling prices of comparable homes with and without a scenic view can be compared to determine the value of the scenic view. Great care must be exercised to determine that the values compared are truly comparable.

Contingent valuation seeks to determine by surveys the value individuals would pay to avoid, or to accept as compensation for, an environmental hazard. Great care must be exercised to eliminate biases in the framing of questions, and even then respondents' answers may be colored by strategic motivations to influence a particular outcome. Nevertheless, for many nonmarket effects, contingent valuation is the only or best means of valuing risks and can be used to value multiple aspects of a complex risk without having to value each aspect separately.

Allocation of Costs to Joint Projects

In allocating costs to multipurpose projects, such as cogeneration or waste-to-energy facilities, the most important initial consideration is whether purposes other than generating electricity dominate the project. If the project would operate regardless of whether it generated electricity—for example, if producing heat were the project's dominant purpose—then none of the project's environmental costs should be allocated to production of electricity. If electricity production is one of the project's main purposes, however, environmental costs can be allocated according to

- the separable costs of fulfilling each purpose
- the value of the product of each process
- the relative importance of the purpose of each process to the plant
- the added emissions from electricity production where calculable, or
- the heat rate of each process

Since environmental costs are being valued, using emissions contributions, if ascertainable, is best. Emissions levels can be measured directly or derived from the amounts of fuel used in producing each product or from the proportional heat rates of each process.

Discount Rates and Real Value Escalation

There is much controversy among economists and other experts about what discount rate, if any, should be applied to environmental externalities. Discount rates are used to compare future economic benefits and costs to today's benefits and costs. Low discount rates weigh the future more heavily (and the present less heavily) than high discount rates do. Thus if a zero discount rate were used, damages to future generations would be counted the same as damages to today's populations. If a 6% discount rate were used, damage to future populations would be decreased 6% each year.

Some maintain that a zero discount rate should be used, particularly for risks to human life and health, because a life in the future is as valuable as a present life (Shuman and Cavanagh 1984). They maintain that sound stewardship of the environment mandates that the value we put on future lives and other environmental assets be considered as highly as present values. Furthermore, they maintain that discounting double-counts the possibility of avoiding future risks since the calculation of present risk already takes into account future events that may diminish the chances of the risk being realized. Lastly, they assert that it is unrealistic to discount long-lasting risks, such as those from high-level nuclear wastes, which pose risks for millennia.

Others reject a zero discount rate for risks to human life and health and argue that the value of damages decreases over time and that a zero discount rate places the present value of future environmental damages much too high. In an extreme case, for example, they assert that few would be willing to pay anything substantial for the risk of human fatality 10,000 years from now: long before that remote time, technology will likely resolve the environmental threat (or the world will be destroyed). In a less extreme example, if forced to choose between a risk of death today and the same risk 50 years from now, it seems likely that the delayed health risk would be preferred (although if the issue were protecting one's own life versus protecting the lives of one's children, an individual might well choose the latter). Use of discount rates higher than zero takes into account the lesser values that may be put on future risks. Placing lower values on future lives than on present ones, they assert, takes into account the lesser willingness of the public to pay for damages far into the future and the likelihood that the risks will be alleviated during long time periods.

Some experts maintain that as a matter of consistency the discount rate applied by utilities to their capital investments (approximately 12%) should be applied to environmental risks. They also argue that using lower discount rates will place the present value of environmental risks too low to have a meaningful influence on resource selection (Chernick and Caverhill 1989).

Many economists adopt a middle ground, using a "social rate of time preference" discount rate (social discount rate), usually in the neighborhood of 3%—lower than utility investment rates, but higher than a zero discount rate. The social rate of time preference is the rate at which society is willing to exchange consumption now for consumption in the future. It reflects the ability of society to remedy environmental hazards over time. The main reason asserted for using a social discount rate rather than the utility discount rate is that the value of environmental costs and benefits *to the public* is being evaluated (and discounted), not the investments of the utility. For the same reason, the social discount rate is preferable to the opportunity cost of public investment and the consumption rate of interest, since these rates measure the costs and benefits of risks to investors and individuals, not to society. Social discount rates should be calculated from the time environmental risk is created (BPA 1986).[5]

Real value escalation estimates the increases in price that will take place over time in environmental and energy resources due to inflation and increased scarcity of finite resources. Real value escalation must be used in valuing environmental externalities.

Uncertainty

Valuation of all environmental externality costs must deal with a considerable margin of uncertainty. Often wide ranges of costs are advanced in different studies. An example is the enormous range in estimates concerning the probability of nuclear accidents. These uncertainties should be dealt with by showing the full range of cost estimates and the bases for those estimates. Then sensitivity analysis should be applied to select a reasonable point within the ranges of the studies, and rationales for the selection should be made explicit.

Despite the uncertainties, the damage costs of most pollution produced by electric utilities can be estimated. The uncertainties involved usually are no greater than the uncertainty in selecting pollution standards to avoid "significant risk to human health and the environment" or "with a reasonable margin of safety," as prescribed by the environmental protection statutes and upheld by the courts.

The principal problem with using environmental damage costs to determine electric utility externality costs is that too few of these damages have been valued adequately. Major research is vitally needed in this area. Reliable damage figures are important not only

5. Pace 1990 used both a 3% social discount rate and a 12% utility discount rate applied to all damage studies it reviewed.

for incorporating accurate externality costs in utility resource selections but also for imposing accurate pollution fees and pollution control standards.

State Incorporation of Environmental Externalities

Presently Used Methods

The methodologies presently used by states to incorporate environmental externality costs include quantitative, qualitative, rate-of-return, and avoided-cost consideration (see Table 2-1, next page). Some states use more than one method; New York, for example, uses quantitative, rate-of-return, avoided-cost (by statute), collaborative, planning, and bidding consideration.

These methodologies have been implemented by the states in planning, bidding, and other resource selection determinations. Collaborative processes between utilities, state agencies, and intervenors for determination of environmental costs and their application have also been used, but with little success. In addition, several innovative methodologies have been proposed for incorporation of environmental externalities in utility planning and resource selection, but no state has yet adopted these methodologies. Table 2-2 (page 51) shows methods used by the states that have begun to act on this issue and by Washington, D.C., Bonneville Power Administration (BPA), and Northwest Power Planning Council (NWPPC).

Orders for Consideration

Of the remarkable number of 29 state public service commissions (PSCs) or legislatures that have taken some action to incorporate environmental externality costs, 19 have issued orders or passed legislation requiring their utilities to take into account these costs in planning and/or bidding. Two states have such orders pending, meaning that a costing proceeding has been established and hearings are under way or that an Administrative Law Judge decision is pending or has been issued and is awaiting commission action. Eight states are actively considering orders for consideration of environmental externalities, meaning that a commission has explicitly stated that it intends to consider externalities.[6]

---

6. *Orders:* Ariz., Calif., Colo., Idaho, Kan., Mass., Mich., Minn., Nev., N.J, N.Y., Ohio, Ore., Penn., Texas, Vt., Va., Wis. *Statute:* Alaska. *Limited orders with more extensive orders pending:* Calif., Mass., Mich. *Orders pending:* Conn., Iowa. *Orders under consideration:* Hawaii, Maine, Md., Mont., N.C., R.I., Utah, Wash., D.C.

Table 2-1. Status of actions incorporating environmental externality costs.

N = No action
O = Incorporation ordered
P = Incorporation order pending
U = Under consideration

	O	P	U	N
Ala.				X
Alaska[a]	X			
Ariz.	X			
Ark.				X
Calif.[b]	X			
Colo.	X			
Conn.		X		
Del.				X
Fla.				X
Ga.				X
Hawaii			X	
Idaho	X			
Ill.				X
Ind.				X
Iowa		X		
Kan.	X			
Ky.				X
La.				X
Maine			X	
Md.			X	
Mass.[b]	X			
Mich.[b]	X			
Minn.[b]	X			
Miss.				X
Mo.[c]				X
Mont.			X	
Neb.				X

	O	P	U	N
Nev.[a]	X			
N.H.				X
N.J.	X			
N.M				X
N.Y.	X			
N.C.			X	
N.D.				X
Ohio	X			
Okla.				X
Ore.	X			
Penn.	X			
R.I.			X	
S.C.				X
S.D.				X
Tenn.				X
Texas	X			
Utah			X	
Vt.	X			
Va.	X			
Wash.				X
Wash. D.C.			X	
W. Va.				X
Wis.	X			
Wy.				X
BPA[a]	X			
NWPPC[a]	X			

Notes:
[a] Established by legislation
[b] Order issued to consider externalities; implementation pending
[c] Commission has stated that it may consider externalities

Source: Pace 1990

Quantitative Consideration

In quantitative consideration, a commission or utility under commission order establishes dollar values for environmental costs. The values calculated are then added to the cost of resources in the selection process or used in a resource rating system. Some commissions attach a proxy percentage adder to polluting resources, a percentage credit to nonpolluting resources, or both.

Table 2-2. Types of actions to incorporate environmental externality costs.

AC = Avoided-cost consideration
B = Bidding consideration
C = Collaborative action
ED = Environmental dispatch
L = Legislative action
O = Incorporation ordered
P = Incorporation order pending
PL = Planning consideration
QL = Qualitative consideration
QN = Quantitative consideration
ROR = Rate-of-return consideration
U = Under consideration

	QN	QL	ROR	AC	ED	C	PL	B	Comment
Alaska				O			L		
Ariz.		O	O					O	
Calif.	O				U	C*	O		*Collab. ended
Colo.	O*							O*	*By fuel type
Conn.	P		L*			C	P	P	*5% ROR adder
Hawaii							U		
Idaho			O	O					
Iowa				P		C			
Kan.	P		O						
Maine		U					U		
Md.	U					C		U	
Mass.	O*					C	P	P	*DSM evaluations
Mich.	U			O			U		
Minn.*		O					P		*Law caps SO_2
Mont.		U	L*				U		*2% adder for DSM
Nev.	O	L					O		
N.J.	O			O				O	
N.Y.	O		O	L		C	O	O	
Ohio		O					O		
Okla.			O*						*ROR, trash only
Ore.*	O	O					O		*Law caps CO_2
Penn.		O*					O*		*Not implemented
R.I.							U		
Texas		O					O		
Utah							U		
Vt.	O					C	O	P	
Va.				O*					*15% DSM adder to AC
Wash.			L*						*2% ROR law for DSM
Wash. D.C.	U						O		
Wis.	O		O		U		O		
BPA	L						L	U	
NWPPC	L						L		

Note: In the categories listed there is overlap; thus New York, for example, is listed as having quantitative, rate-of-return, avoided-cost (by statute), collaborative, planning, and bidding considerations.

Source: Pace 1990.

As Table 2-2 shows, 13 states and Washington, D.C. have adopted quantification or have quantitative orders pending or under active consideration. Nine states plus the BPA and NWPPC have acted to consider environmental externality costs quantitatively or to use a proxy adder to represent these costs. Two states have such quantification orders pending, and two states and Washington, D.C. have them under active consideration.[7]

The New York PSC has been the pioneer in incorporating quantified environmental externality costs, requiring its utilities to assign about 15% of total bid evaluation scoring points to environmental externality costs (about 24% of price scoring points), calculated at 1.405 ¢/kWh total environmental externality costs, based on a coal-fired plant meeting New Source Performance Standards (NSPS).[8] These same environmental costs must be used in valuing demand-side management (DSM) investments in integrated resource planning.[9] The PSC has also ordered all New York utilities to do a pooled study, with participation of outside experts and the public, to quantify the environmental externality costs of pollution from their utilities' operations.[10]

In Wisconsin, a noncombustion credit/adder has been adopted for screening of all utility resource acquisitions, so that a noncombustion source that costs 15% more than a combustion source will be considered on a par with the latter; this screening is followed by a qualitative test requiring consideration of environmental impacts.[11]

---

7. *Quantified orders:* Calif., Colo., Mass., Nev., N.J., N.Y., Ore., Vt., Wis. *Limited order with more extensive order pending:* Calif. *Quantified orders pending:* Conn., Mass. *Under consideration:* Md., Mich., Wash. D.C.

8. See N.Y. PSC Case 88-E-241, *Proceeding on Motion of the Commission (established in Opinion 88-15) as to the guidelines for bidding to meet future electric capacity needs of Orange & Rockland Utilities, Inc.,* Order Issuing a Final Environmental Impact Statement and Adopting Staff's Response to Agency Comments (24 March 1989).

9. *Formats and Guidelines for July 23, 1990, DSM Plan Filing in Case 29223.* N.Y. Department of Public Service Staff (23 February 1990). The guidelines provide that

> Environmental benefits are to be explicitly quantified in the total resource cost test. The Staff estimates of environmental costs, developed initially in the electric capacity bidding cases, should be used in the assessments for the 23 July, 1990 plan.
> The environmental benefits to be used are
> 1.4¢/kwh for programs that promote energy efficiency
> 0.9¢/kwh for programs that are aimed at peak clipping
> 0.4¢/kwh for programs that are aimed at load shifting

10. NY PSC Case 28223, *Electric Utility Conservation Programs,* Opinion and Order 89-15 (23 May 1989).

11. Wis. PUC *Re Advance Plans,* Docket 05-EP-5, 102 P.U.R. 4th 245 (6 April 1989).

Also, in integrated planning, resource valuations must assume that carbon dioxide emissions will have to be reduced to 80% of their 1985 level by 2000 and to 50% of the 1985 level in the long run.[12] Wisconsin legislation concerning acid rain also requires utilities to cut 1980 levels of sulphur dioxide emissions by 50% by 1993 (Cohen and Eto 1989); while not directly used in valuation, this statute is relevant in that it internalizes some of the costs of acid rain.

The Oregon commission has not quantified environmental externality damage costs but has ordered the utilities to do so "to the fullest extent practicable" and to indicate ranges of costs where definite damage costs cannot be determined." The utilities are required to consider these damage costs in resource selection.[13]

The Massachusetts Department of Public Utilities has just issued an order that values environmental externalities for an NSPS coal plant about 4¢/kWh.[14] The California Energy Commission has set a similar value on NO_x power plant pollution, and the California PUC is completing a proceeding requiring that these values be incorporated in utility resource selection.

Qualitative Consideration

Nine states presently have ordered, or are considering ordering, that their utilities take into account environmental externality costs—in planning or resource selection or both—without specifying how these costs are to be calculated or considered.[15]

Rate-of-Return Consideration

Rate-of-return consideration involves an award by a commission of an increased rate of return to utilities as an incentive to install nonpolluting resources. Increased rate of return is typically allowed either on particular nonpolluting resource investments—usually DSM, renewables, or resource recovery plants—or on total investments. While DSM incentives are adopted primarily to make DSM investments as profitable as supply investments, not necessarily to capture environmental externality costs, most commissions cite environmental benefits as one of the reasons for offering such incentives. We have included in this compilation the states where this is so.

12. Wis. Stat. Ann. Sec. 144.385-389 (1989).
13. Ore. PSC, *Re Least Cost Planning,* UM 180, Order No. 89-507 (20 April 1989).
14. Mass. DPU Order 89-239 (31 August 1990).
15. *Orders:* Ariz., Minn., Nev., Ohio, Ore., Penn., Texas. *Under consideration:* Maine, Mo. Wisconsin uses qualitative consideration after applying a 15% noncombustion credit/adder (see footnote 11).

As Table 2-2 shows, nine states give rate-of-return consideration to environmental externality costs; Oklahoma does so for resource recovery plants only.[16]

Washington and Montana have statutes providing a 2% additional rate of return on energy conservation investments and Connecticut a 5% DSM rate-of-return adder, citing environmental externality costs as a justification. Kansas gives a 0.5%–2% increased rate of return on renewable and conservation resources. The Idaho commission has announced that in future rate cases it will take into account utility conservation efforts in determining the allowed rate of return on total investments. The Wisconsin commission experimentally has allowed Wisconsin Electric Power Co. to earn an additional 1% rate of return for each 125-MW reduction in peak load the utility achieves through efficiency investments. New York has adopted, as a temporary DSM incentive, the return of lost revenues from DSM investments plus a performance-based incentive.

Avoided-Cost Consideration

Seven state legislatures or commissions have required, or are considering requiring, adding a premium to utility-calculated avoided power plant capacity and energy costs, to help account for environmental externalities.[17] Thus, New Jersey has established avoided cost under the Public Utility Regulatory Policies Act of 1978 (PURPA)[18] at 10% over the regional power pool's energy billing rate to reflect the potential cost savings to society from the presumably more environmentally benign "qualifying facilities" (as defined in PURPA). The Virginia Commission requires addition of 15% to a utility's avoided-cost submission, also based on societal costs.

The New York State Legislature, citing environmental considerations, set a statutory 6¢/kWh PURPA avoided-cost rate, which is above most utility-calculated avoided costs; the Federal Energy Regulatory Commission (FERC) voided application of this rate, interpreting PURPA to prohibit reimbursement in excess of avoided cost. Many state commissions objected to this decision as an unwarranted usurpation of state rights. The FERC decision is being appealed, and one of the FERC commissioners has stated that application of the decision to other situations would be decided on a generic basis.

Collaborative Consideration

Six states are involved in collaborative efforts among utilities, state

16. *Orders:* Idaho, Kan., N.J., N.Y., Okla., Wis. *Statutes:* Conn., Mont., Wash.
17. *Ordered:* Alaska, Idaho, Mich., N.J., N.Y., Va. *Pending:* Iowa.
18. 16 U.S.C. 2601 et. seq.

regulators, and intervenors to determine how environmental externality costs will be calculated and incorporated.[19] None of these efforts has yet come to fruition, and a recent collaborative DSM effort in California resulted in inability of the parties to agree on externality values or incorporation methodologies (CEC 1990).

Planning Consideration

Twenty-two states, Washington, D.C., and the BPA and NWPPC require or are contemplating requiring consideration of environmental externality costs in least-cost planning; twelve states now have requirements, and eleven states have requirements pending or under consideration.[20]

Bid Evaluation Consideration

Eight states and BPA require, are considering requiring, or have orders pending to take into account environmental externality costs in evaluating bid scores.[21] Three state commissions—New York's, New Jersey's, and Colorado's—have provided for specific points to be assigned in bid evaluations to account for environmental externality costs, but only New York's effort is substantial.

Proposed Incorporation Methods

Environmental Dispatch

In environmental dispatch, a commission orders a utility or power pool either to dispatch environmentally benign resources ahead of more polluting resources—even though the latter may cost less—or to dispatch resources according to least cost, with environmental costs included in the determination of least cost. Environmental dispatch alleviates environmental damages and costs while displacing production from the most heavily polluting power plants and thus encouraging early closure of those plants.

No states currently are employing environmental dispatch as a means of incorporating environmental externality costs. All the methods presently adopted by states to incorporate environmental externality costs address only resource selection to meet new capacity or new energy needs. In its mandated bidding regime, however, the New York commission requires utilities to include life extension of existing

19. Conn., Iowa, Md., Mass., N.Y., Vt.

20. *Orders:* Ariz., Calif., Nev., N.Y., Ohio, Ore., Penn., Texas, Wash. D.C., Wis. *Statute:* Alaska, Vt. *Orders pending:* Conn., Mass., Minn. *Orders under consideration:* Hawaii, Maine, Mich., Mo., Mont., N.H., R.I., Utah (statute).

21. *Orders:* Colo. (by fuel type), N.J., N.Y. *Pending:* Conn., Mass., Vt. *Under consideration:* Md.

plants as a resource option and to determine the appropriateness of continuing to operate those plants by comparing their costs with the prices bid for new resources. Wisconsin's recently enacted acid rain statute includes environmental dispatch among the compliance options for meeting the SO_2 and NO_x standards.[22]

The Ohio Office of Consumers' Counsel did a recent study on cleaning up the state's very substantial contribution of acid rain precursors, finding that a combination of least-emissions dispatching and aggressive investment in energy efficiency could prevent increases in SO_2, reduce cleanup costs by more than 60%, and reduce cumulative costs for electric energy services by as much as $3 billion through 2005 (Centolella 1988). A model has been developed for analyzing the environmental costs and benefits of environmental dispatch (Heslin and Hobbs 1989). BPA's production models and resources planning models are capable of analyzing environmental dispatch but are usually run to determine the lowest social costs of meeting utility system loads. The Cornell Carnegie-Mellon model developed to model New York utility emission impacts could also accommodate environmental dispatch (Pace 1990).

Ranking

The Center for Global Change at the University of Maryland has worked out an innovative ranking and weighting methodology for evaluating environmental externality costs.[23] The problem with all ranking systems, however, is that their accuracy must inevitably be judged on the degree to which they approximate costs. To the extent that they depart from costs, they produce significant ranking and cost distortions. Using the best cost data available is easier to understand and can be varied more readily as new cost data become available.

"Environmental LCUP"

"Environmental LCUP (Least Cost Utility Planning)" is an innovative concept proposed by Florentin Krause of Lawrence Berkeley Laboratory for incorporating environmental externalities. Under the proposal, emission reduction targets would be set for principal power plant pollutants like CO_2, SO_2, NO_x, and particulates, and the utilities would be required to meet these targets in a least-cost manner. As an enforcement mechanism, utilities could receive a

22. Wis. Stats. Ann., Secs. 144.385–144.387; see also Secs. 15.347, 16.02.
23. Vermont Public Service Board, *Application of Twenty-four Electric Utilities . . . for a Certificate of Public Good Authorizing Execution and Performance of a Firm Power and Energy Contract with Hydro-Quebec and a Hydro-Quebec Participation Agreement,* Docket No. 5330, testimony of Susan Hedman, Alan S. Miller, and Irving Mintzer (January 11, 1990).

positive rate-of-return incentive for meeting or exceeding the targets and a negative incentive for failing to meet them. Wisconsin, with its new acid rain law, comes as close as any state has to using this proposed method of setting emission standards that utilities would have to meet at least cost.

The advantage of environmental LCUP is that it avoids the necessity for calculating environmental externality costs and is designed to achieve specific emission reduction targets. It lends itself well to valuing hard-to-calculate regional and global externalities. The disadvantage is that it requires the setting of emission reduction targets, which may be as difficult as environmental externality costs for commissions to calculate. Their doing so also may be viewed as invading legislative prerogatives. However, environmental LCUP could be used for hard-to-value resources only, with more readily established values used for other resources. Environmental LCUP might be used for CO_2 costing—requiring, for example, a 20% reduction in a least-cost manner—and conventional methods used for valuing other pollutants.

Assessment of Environmental Costs Against Resources and Creation of a Pollution Mitigation Fund

An innovative proposal by former Maine Public Utilities Commissioner David Moskovitz would charge resource owners with the quantified environmental costs of each resource selected and deposit the proceeds in a pollution mitigation fund, thus internalizing the environmental costs. This proposal has the enormous advantage of making resource owners pay the costs of the environmental damages they impose on society instead of just using these costs in resource selection. It would also create a very substantial fund that could be used for environmental mitigation and for promotion of use of environmentally benign renewable resources and marginally cost-effective DSM programs. Instituting such a system may be beyond the statutory authority of many commissions, although the New York commission did require its utilities to devote 0.25% of gross revenues to establish a fund for DSM research and experimentation, including the assessment of environmental costs.[24]

Recommendations for Incorporation

There has not yet been sufficient experience with incorporating

---

24. N.Y. PSC, Case 28223, Opinion and Order 84-15, *Requiring the Development of Conservation Programs,* 21 May 1984.

environmental externality costs under any of the statutes or state commission orders described above to be able to ascertain which methodology will work best. Considering all the pros and cons of the various proposals, we propose the following recommendations:

1. Environmental externality costs should be incorporated in all utility planning, bidding, and other resource selection.
2. Quantified environmental externality costs should be used, based on damage costs where adequate valuation studies are available, and otherwise based on control costs.
3. A major research effort is critically important to better determine environmental damage costs.
4. Rate-of-return incentives should be provided for acquisition of energy-efficient resources, such that a kWh saved will be as profitable as a kWh sold; this step may require decoupling profits from sales.
5. Environmental externalities should be included in setting avoided costs.
6. Environmental costs should be internalized by an assessment against resources selected and should be placed in a pollution mitigation fund.
7. Environmental dispatch and environmental LCUP should be tested to determine their environmental and ratepayer effects and applicability in cases of hard-to-quantify pollution costs.

Next Steps

Environmental externality valuation is still at an early stage of development. Much research is needed to get firm and defensible costing figures. The U.S. Department of Energy and the U.S. Environmental Protection Agency should perform a thorough study of quantifying environmental damage costs, on a scale comparable to the congressionally mandated National Acid Precipitation Assessment Program (NAPAP) study of acid rain impacts and damages. The research area requiring greatest attention is dose-response relationships.

While 29 state jurisdictions and Washington, D.C., the BPA, and NWPPC have started to consider environmental externalities, many of their efforts are tentative. A great deal of experimentation is needed on the various means of incorporation that have been attempted and proposed. A concerted effort should be made to exchange information among state commissions and utilities and to prompt other commissions to incorporate environmental externalities.

It is heartening that the National Association of Regulatory Utility Commissioners has taken a major interest in this area, having held a national conference on the subject in October 1990 in Jackson Hole, Wyoming.

Information must also be exchanged among environmental costing experts, a process requiring a unique collaboration of economists, scientists, and utility experts. In fall 1990, the Pace University Center for Environmental Legal Studies and Fraunhofer Institut held an international costing conference sponsored by the German Marshall Fund of the United States and the Daimler-Stiftung Foundation. Academic, utility (Electric Power Research Institute, Gas Research Institute), and government research institutes should devote major efforts to both quantification and incorporation issues.

Acknowledgements

This paper is based on a report prepared by the Pace University Center for Environmental Legal Studies for the New York State Energy Research and Development Authority, supplemented by funding from the U.S. Department of Energy and the National Audubon Society (Pace 1990). The report was prepared by Pace Center staff; Pace University Law School students; and consultants Shepard C. Buchanan of Bonneville Power Administration, Paul Chernick and Emily Caverhill of PLC, Inc., and Alan Krupnick of Resources for the Future.

References

Bland, P. 1986. "Problems of Price and Transportation: Two Proposals to Encourage Competition from Alternative Energy Resources." *Harvard Law Review* 10: 2.

Bonneville Power Administration. 1986. "Cost-Effectiveness Methodology." In *Bonneville Power Administration Resource Strategy.* Report DOE/BP/751. Portland, Ore.: Bonneville Power Administration.

BPA. See Bonneville Power Administration.

Brick, S. 1989. "Strategies and Methods for Incorporating Environmental Concerns Into Utility Planning and Utility Regulatory Decision-Making." Middleton, Wis.: MSB Energy Associates.

California Energy Commission et al. 1990. *An Energy Efficiency Blueprint for California: Report of the Statewide Collaborative Process.* Sacramento, Calif.: California Energy Commission.

CEC. See California Energy Commission.

Centolella, P. 1988. "Clearing the Air: Using Energy Conservation to Reduce Acid Rain Compliance Costs in Ohio." Columbus, Ohio: Ohio Office of the Consumers' Counsel.

Chernick, P., and E. Caverhill. 1989. "The Valuation of Externalities from Energy Production, Delivery, and Use." Boston: PLC, Inc. Photocopy.
Cohen, S. D., and J. Eto. 1989. "Preliminary Results of LBL/NARUC Externalities Survey." Berkeley, Calif.: Lawrence Berkeley Laboratory.
Economist. 1989. "Money from Greenery." *The Economist,* 21 October 1988.
Electric Power Research Institute. 1988. *Benefits of Environmental Controls: Measures, Methods and Applications.* Palo Alto, Calif.: Electric Power Research Institute.
EPRI. See Electric Power Research Institute.
Freeman, A. 1979. *The Benefits of Environmental Improvement: Theory and Practice.* Baltimore: Johns Hopkins Press.
Heslin, J., and B. Hobbs. 1989. "A Multiobjective Production Costing Model for Analyzing Emissions Dispatching and Fuel Switching." *IEEE Transactions on Power Systems* 4 (3).
Hill, R. 1990. "The Impact of Energy on Environment and Development." In *IVth Nobel Prizewinners Meeting, December 1989, Man, Environment, and Development—Towards a Global Approach.* Nova Spes Meeting Proceedings.
Hohmeyer, O. 1989. *Social Costs of Energy Consumption.* Berlin: Springer-Verlag. (Prepared under contract for the Commission of the European Communities, Directorate-General for Science, Research and Development, Document No. EUR 11519, by Fraunhofer-Institut fur Systemtechnik und Innovationsforschung, Karlsruhe, Federal Republic of Germany.)
Machado, S., and R. Piltz. 1988. "Reducing the Rate of Global Warming: The States' Role." Washington, D.C.: Renew America.
Mitchell, R., and R. Carson. 1989. "Using Surveys to Value Public Goods: The Contingent Valuation Method." Washington, D.C.: Resources for the Future.
NAPAP. See National Acid Precipitation Assessment Program.
National Acid Precipitation Assessment Program. 1987. *Interim Assessment.* Vol. 1.
National Regulatory Research Institute. 1982. "The Appropriateness and Feasibility of Various Methods of Calculating Avoided Costs."
Pace. See Pace University Center for Environmental Legal Studies.
Pace University Center for Environmental Legal Studies. 1990. *Environmental Costs of Electricity.* Report for the New York State Energy Research & Development Authority. Dobbs Ferry, N.Y.: Oceana Publications.
Reid, M. W. 1988. *Ratebasing of Utility Conservation and Load Management Programs: Final Report.* Report prepared for Northeast Utilities. Washington, D.C.: Alliance to Save Energy.
Shuman, M., and R. Cavanagh. 1984. *A Model Electric Power and Conservation Plan for the Pacific Northwest, Environmental Costs.* Portland, Ore.: Bonneville Power Administration.
Vanderlindin, B. 1988. "Bidding Farewell to the Social Costs of Electricity Production: Pricing Alternative Energy under the Public Utility Regulatory Policies Act." *Journal of Corporate Law.*

Chapter 3

Electric Utility Conservation Programs: A Review of the Lessons Taught by a Decade of Program Experience

Steven Nadel, *American Council for an Energy-Efficient Economy*

In many regions of the country, electric utilities are now the leading sponsors of energy conservation programs, having surpassed government (whose activities have declined due to budget cutbacks) and the private market (where the decline in real energy prices has slowed conservation efforts). More than 500 utilities have sponsored demand-side management (DSM) programs including approximately 1,000 residential programs (Blevins and Miller 1989a) and 400 commercial and industrial (C&I) programs (Blevins and Miller 1989b). Total expenditures on electric utility DSM programs now exceed $1.2 billion annually (Tempchin et al. 1990). Utility program offerings are particularly numerous on the West Coast and in the northeastern and north central United States.

This chapter examines the basic types of residential and commercial conservation programs that have been offered, including information, rebate, loan, performance contracting, comprehensive direct installation, and bidding programs. DSM program types not discussed include load management programs (reducing peak demand and/or shifting energy demand from one time period to another) and load building programs (increasing loads during some or all periods in order to meet a utility's strategic objectives). This chapter reviews typical and exemplary programs and provides information on participation rates, free riders, energy savings, cost per unit of energy savings, and factors linked with program success (as available data permit). Drawing on these results, three final sections (1) summarize key lessons learned, (2) put overall program results in perspective, and (3) discuss issues and directions for the 1990s.

This chapter focuses particularly on programs and program procedures that are cost-effective to the sponsoring utility and that result in high net participation rates and/or high net electricity savings (net of what would have happened if the program was not offered). If demand-side resources are to play a major role in meeting future electricity needs, then programs will need to reach a substantial proportion of customers and will need to have a significant impact on the electricity consumption of the customers that are reached. If high participation is achieved but savings per customer are minimal, or if high savings per customer are achieved but participation is low, then the total savings achieved will be limited. For example, if a utility C&I rebate program reaches customers responsible for 20% of total energy sales and reduces energy use by these customers by 7%, then C&I energy use will be reduced by 1.4% (20% times 7%). While a reduction in C&I energy sales of this magnitude is significant, it will have little impact on a utility's long-term need for generating plants.

Data on programs examined in this chapter come from recent survey reports on residential (Schick et al. 1990) and C&I (Nadel 1990a) programs. Additional data come from reports by, and conversations with, the individual utilities operating the programs. The ways different types of data are defined and tracked vary considerably among utilities. For example, some utilities track only direct program costs such as rebates paid to customers, others include some indirect costs such as advertising costs, and still others try to account for all indirect costs including staff costs and program planning and evaluation costs. (This issue is discussed extensively in Berry 1989.) Due to these variations, comparisons between programs are subject to a considerable margin of error. In order to prevent misinterpretation, the following definitions and limitations apply unless otherwise stated:

Participation rate is the cumulative number of customers participating in a program divided by the number of customers eligible for the program. Participation rates reported here generally include free riders (customers who participate in a program but would have implemented a conservation measure even if no program were offered). In the discussions, allowance is made for free riders, to the extent available data permit. In interpreting participation-rate figures, bear in mind that the more limited and targeted the eligible population, the easier it is to achieve high participation rates, but these high rates may come at the expense of leaving some customers ineligible for a program.

Electricity savings are reported as a percent of the average participating customer's preprogram electricity use. Wherever possible, savings figures reported in this chapter are based on a statistical comparison of electricity bills for program participants and nonparticipants. These savings estimates are referred to as "net savings" because savings are net of a control group of nonparticipants. For most programs, net savings figures are unavailable. In these cases, either "gross savings" (savings determined with a billing analysis that does not include a control group) or engineering estimates of savings are used, and are so indicated in the text.

Costs per kWh are based on utility costs (including indirect costs such as staff and marketing) and do not include costs borne by the customer. Unless otherwise stated, costs per kWh are levelized using the California Standard Practice Approach (CPUC 1987) over the assumed life of the measure and using a 6% real discount rate. Cost per kWh figures are only approximate; they use simple analysis procedure and sometimes rely on rough estimates of indirect costs. Two other common measures of program cost effectiveness, cost per kW and societal cost per kWh, are not reported here. Cost per kW figures are difficult to compare because utilities vary as to whether and how they adjust kW savings for coincidence with the system peak. Societal costs for a program include utility costs, customer costs not paid by the utility, and, sometimes, the cost of externalities, such as pollution, associated with a program. Societal costs are not reported here because, at present, few utilities collect this information.

Residential Programs

Nationwide, residential electricity use accounts for approximately 35% of kWh sales (EIA 1989b). The residential sector is made up of many small customers who have relatively homogeneous energy consumption patterns (relative to the C&I sectors). Nationwide, approximately 20% of residential electricity use is for refrigerators and freezers, 19% for space heating, 17% for water heating, 13% for air-conditioning, 11% for lighting, and 20% for other uses (BNL 1987). End-use allocations for space heating and air-conditioning vary widely depending on local climatic conditions and on local saturation rates for electric heating and cooling.

Residential conservation programs fall into three general categories: retrofit programs (improvements to existing homes and equipment), new construction programs, and new equipment programs (promoting high-efficiency equipment at the time of equipment replacement). Each of these categories is discussed below.

Retrofit Programs

Residential retrofit programs span a continuum from audit-only programs (where the utility provides information, but the customer provides financing and arranges for measure installation) to comprehensive programs where the utility provides information, financing, and installation for all cost-effective measures. There are many ways to categorize programs along this continuum; no single approach is widely used. In this chapter we use five categories: audit programs; other information programs; low-cost-measure programs such as lighting, water heating, and low-cost weatherization; moderate-cost-measure programs such as attic insulation and infiltration reduction; and comprehensive programs such as complete weatherization packages. In addition, special programs aimed at multifamily buildings and low-income customers are briefly discussed.

Audits

Most electric and gas utilities offered residential energy audits during the 1980s as part of the federally mandated Residential Conservation Service (RCS) program. These programs generally provided on-site computerized energy audits and supplied a detailed audit report to the customer. Audits were generally provided to customers for a nominal charge (typically $10), although some programs offered free audits. According to an evaluation of the program six years after its beginning, approximately 7% of eligible customers had participated in the program (DOE 1987). At least six statewide programs and ten utility programs had achieved participation rates in excess of 15%, including one statewide program with a 25% rate and three small-utility programs with rates in excess of 35%. Factors linked with high participation rates included state and utility commitment, the provision of financing assistance, and assistance helping customers arrange for measure installation.

Evaluations of the program found audited households had average net savings of 3–5% (Hirst 1984). These programs did not include any financial incentives or arranging assistance. Several studies (for example, Hirst and Hu 1983) found the program cost-effective from the utility and societal perspectives when savings from all fuels were considered, although net benefits from the program were small. We can surmise from this finding that audits on electrically heated homes are probably cost-effective to an electric utility, but providing a complete audit on a gas-heated or oil-heated home is unlikely to be cost-effective to an electric utility.

In recent years, a number of new audit programs have been developed that seek to reduce audit complexity and costs while increasing implementation of audit recommendations. For example, Boston Edison's EASY program uses a simplified audit procedure and a small portable computer to generate an audit report while the auditor is still in the home. The auditor then explains the audit results and highlights the most important recommendations. In addition, at the time of the audit, the auditor installs up to $25 worth of free energy conservation materials, such as a water heater wrap, weather stripping, and a compact fluorescent bulb. Local community organizations under contract to the utility perform the audits. Savings from the program have not been evaluated (Patricia McCarthy, Boston Edison Co., personal communication).

Other Information Programs

Data on residential information programs are limited, so assessing the effectiveness of these programs is difficult. The limited data available tend to indicate that simple information efforts such as brochures have a limited impact, but that other informational programs such as home energy rating systems (HERS) and energy use feedback programs may be more promising. For example, Collins and associates (1985), in a study on evaluation results, examined a number of simple information programs including information pamphlets, hotlines, videos, and appeals to conserve. Most of these programs achieved net or gross energy savings of 0–2%.

HERS programs rate homes on their level of energy efficiency, under the premise that consumers will strive to achieve high ratings for their homes. For example, the Austin Energy Star program rates homes on a zero to three scale based on the results of an energy audit (Vories 1986). While some HERS programs cover existing homes, most programs have emphasized new homes. Among existing homes, participation rates of up to 20% have been reported (Vine, Barnes, and Ritschard 1987); data on energy savings for existing homes are unavailable.

Feedback programs provide monthly or annual feedback to consumers on their electricity use relative to previous years, to a typical household, or to both. This feedback can have two effects: those consumers with above-average use may be motivated to take action; and all consumers, seeing the impacts of their actions, may be motivated to take additional actions. A number of short-term, small-scale research studies have shown that feedback programs can have a significant impact on consumer energy use (summarized in Collins et al. 1985). On a utility-wide scale, National Fuel, a gas utility, mailed annual energy use reports to all its residential customers, and as a result, requests for energy audits increased dramatically. The annual cost of the program was $.32 per customer. No effort was made to measure energy savings attributable to the program (W. Kempton, Princeton University, personal communication).

Low-Cost-Measure Programs

In low-cost-measure programs, participation rates can be high if customer costs are kept low, but energy savings per customer are limited by the small number of measures promoted. For example, in an experimental program, Niagara Mohawk offered customers water heater blankets, pipe wrap, low-flow shower heads, and compact fluorescent lamps under three different financing arrangements: free, at half cost, and at full cost but financed through a deferred payment loan. Installation was generally provided by the utility. Participation rates averaged 44% for the free offer, 6% for the half-cost offer, and 3% for the loan (Research Triangle Institute 1989). Information on program savings and cost-effectiveness are not available.

Similarly, under Central Maine Power's (CMP) Packaged Weatherization Services program, private contractors hired by the utility installed a package of low-cost measures—including caulking, weather stripping, and interior storm windows—in a customer's home for a nominal fee. The program was marketed via direct mail. The contractor billed the utility for the difference between the actual average cost of $248 and the fee paid by the customer. Three fee levels were tested: $19.95, $49.95, and $79.95; participation rates were 25%, 12%, and 7% respectively. Net savings from the program relative to a control group of nonparticipants averaged 3%. Measured energy savings averaged 36% of savings estimated by previous engineering studies. The low savings were attributed to use of secondary fuels such as wood in the participating homes (the engineering estimates assumed all the savings would be taken in the form of electricity, when in actuality, much of the savings showed up as reduced wood use). Due to the disappointing level of actual savings, utility costs for the program averaged $.098/kWh (Ecker and Michelsen 1989). Assumptions used to calculate the cost per kWh differed slightly from the California Standard Practice Approach.

Innovative approaches may be able to achieve participation

rates over 40%. For example, the city of Santa Monica, California, and the Michigan Public Service Commission have achieved participation rates of 33–35% and 36–59% respectively with an Energy Fitness program (Kushler, Witte, and Ehlke 1989). In the Energy Fitness program, low-cost measures are provided and installed for free. Marketing is done by neighborhood blitz, in which advance publicity lets homeowners know when the program will be in their neighborhoods, and door-to-door canvassing is used to solicit participation. Measures are installed on the same day that a home is canvassed. The Santa Monica program primarily served homes with gas space-heating and water-heating. Net gas savings due to the program averaged 5.4%. The simple payback period for the program, including all program costs, was estimated to be less than 3.5 years (Egel 1986).

The Santa Monica and Michigan programs offered measures that saved both electricity and fossil fuels. However, the New England Electric System (NEES) has recently used the same approach to promote low-cost electricity conservation measures including compact fluorescent lamps, basic maintenance on refrigerators and air-conditioners, and low-cost water-heating and space-heating conservation measures for customers who use electricity for these end uses. Preliminary results are that 44% of eligible customers are participating in the program (about 70% of customers contacted) (Betty Tolkin, NEES, personal communication). Estimated annual savings, based on engineering estimates, are 555 kWh/participant; percentage savings are not presently available (NEES 1990). Based on this savings estimate, and assuming a six-year average measure life, program costs averaged $.041/kWh.

Programs to increase the efficiency of water heating have also achieved high participation rates. In these programs a utility staff person or contractor visits the homes of electric-water-heating customers to install a water-heater wrap, pipe insulation, low-flow shower heads, and aerators. In some programs, installers also reduce the water-heater thermostat setting if the homeowner so permits. NEES achieved a participation rate of 65% with its program after 2.5 years, not counting an additional 10% of customers who had installed a wrap prior to the program (Dick Morency, NEES, personal communication). The NEES program was promoted through extensive telemarketing. Services were provided at no charge to the customer, and installation was scheduled at times convenient to the customer, including evenings and weekends. Net annual savings from the program were estimated to average 533 kWh/customer for water-heating measures plus an additional 64 kWh/customer for four reduced-wattage incandescent lightbulbs (NEES 1990). Assuming a six-year measure life, program costs average $.02/kWh saved. Similarly, Seattle City Light achieved a 62% participation rate with a water-heater wrap and thermostat turn-down program, which was marketed door to door, neighborhood by neighborhood (T. Newcomb, Seattle City Light, personal communication).

Moderate-Cost-Measure Programs

Programs promoting moderate-cost measures such as attic insulation and infiltration reduction measures have generally not been as successful as their designers had hoped, although efforts to develop successful approaches continue. Experiences to date tend to indicate that without extensive utility assistance with packaging, arranging, and financing measures, participation and savings are likely to be low. For example, CMP's Energy Management Rebate Program offered rebates of 15–50% for conservation measures chosen by customers. Technical assistance from the utility was limited to a standard RCS audit. Participation rates ranged from 1% to 2%, with the amount of the rebate having little impact on customer response. Net savings averaged 2%, significantly less than engineering estimates had predicted. Again, the disappointing actual savings are attributed to use of secondary fuels (Ecker and Michelsen 1989).

Similarly, Massachusetts Electric's Enterprise Zone program featured free energy audits with a menu of low-cost and moderate-cost conservation measures available without subsidy. The utility arranged measure installation. The program was heavily marketed through mailings, local events, and telemarketing. Participation rates were 26% for the audit but only 6% for the measure installation service. Due to high marketing costs and low installation rates, utility costs averaged $.067/kWh saved (assuming a ten-year measure life and based on engineering estimates of savings). Interestingly, a variation on the program that offered free conservation measures to low-income customers cost the utility only $.028/kWh. Marketing costs were substantially lower for the low-income group of customers, and participation rates and savings per customer were substantially higher (NEES1988a).

A number of utilities have used a performance contracting approach to promote moderate cost measures. Under performance contracting, private contractors are paid for each kWh they save based on an analysis of customer bills before and after program participation. The private contractors market the programs and install and finance the conservation measures. In most cases there is no cost to the customer, although a few programs include limited customer charges. Typically, the programs offer measures to reduce both water-heating costs and infiltration.

In general, residential performance contracting results indicate that these programs can achieve substantial participation but that actual savings have been significantly less than preprogram estimates provided by performance contracting firms. For example, General Public Utilities' (GPU) RECAP program served approximately 40% of targeted customers (all-electric homes whose electricity consumption exceeded a minimal level) using a combination of mail, phone, and word-of-mouth marketing within specific communities. The average cost per home (including marketing, labor, and other indirect costs) was $670, and net savings averaged 6.6%, considerably less than the 15–30% savings estimated at program conception. Reasons for the discrepancy were not provided in the evaluation (Brown and Reeves 1985; Brown and White 1987). Similarly, Western Massachusetts Electric Company's (WMECO) Performance Contracting Pilot Program had a participation rate of approximately 42% of eligible households (customers with electric water-heating, electric space-heating, or both) including 54% of low-income households. Retrofit costs averaged $200–$250/customer, and net savings averaged 2.7%, considerably less than the 12% previously estimated by the contractor. The savings were less than original estimates due partly to problems with material quality and workmanship (Temple, Barker, and Sloane 1987; Peach 1989). Due to the disappointing savings results, utility costs for the GPU and WMECO programs were $.067/kWh and $.083/kWh respectively, assuming a ten-year average measure life.

Comprehensive Programs

Comprehensive programs generally combine audits with financial incentives. Assistance arranging for measure installation is sometimes provided as well. When financial incentives are high, substantial participation and savings rates can be achieved. For example, Bonneville Power Administration's (BPA) Residential Weatherization program offers customers with electric space-heating a free energy audit, arranges for measure installation (primarily insulation and storm windows), and provides a rebate averaging 75% of measure costs (rebates are based on engineering estimates of the annual kWh savings). In the first six years of program operation, 23% of eligible customers received rebates. Due to the high rebate amount, the program has been popular with customers, so only limited marketing has been needed. Participation has been limited by available budgets (Schick et al. 1990). A variation on the program was offered in Hood River, Oregon, where conservation improvements were installed in 85% of eligible homes by combining 100% grants with intensive, community-based marketing (Hirst 1987).

Depending on the year and program variation being analyzed, net savings for BPA's weatherization programs have averaged approximately 12% (Schweitzer et al. 1989). Actual savings averaged 53% of savings predicted by the energy audits. Reasons for the discrepancy include use of supplemental heating fuels and other building-occupant interactions, such as rooms that are not heated, that were not modeled in the audit (Ken Keating, BPA, personal communication). Utility costs for the basic weatherization program have averaged $.031/kWh or $.040/kWh if customer costs are included (Schick et al. 1990). In calculating these costs, a 3% real discount rate and a 31-year measure life are assumed. If a 6% discount rate and a 20-year measure life are used instead (as is done for other programs discussed in this chapter), utility and societal costs climb to $.054/kWh and $.07/kWh respectively. Schick and associates (1990) report that BPA is planning to lower the incentive level, although not so low as to affect participation significantly.

High participation and savings have also been achieved with comprehensive programs that feature loans instead of grants. For example, the Tennessee Valley Authority (TVA) Home Weatherization program, which ran from 1977 to 1989, combined audits with contractor referrals and zero-interest loans. In the latter years of the program, interest was charged on loan amounts greater than $1,200. Over the 12 years, 62% of electrically heated homes were weatherized under the program. Many techniques were used to market the program, but the most effective techniques were newspaper ads, bill stuffers, and cooperative advertising with trade allies. The average investment per home was approximately $1,000. Insulation, storm windows, caulking, and weather stripping were the most common measures. Annual savings were estimated at 3,000 kWh/home, and utility costs averaged $.02/kWh saved; details on the assumptions used to make these calculations are not available (Schick et al. 1990).

The relative merits of loans versus grants have been debated considerably. Available research tends to indicate that zero-interest loans result in higher participation rates than low-interest loans. In comparing zero-interest loans to grants of equivalent cost to the utility, studies tend to indicate that the majority of customers prefer grants, particularly customers from lower income groups, although a significant minority (ranging from less than 15% to as many as 49%, depending on the study) prefer loans (Stern, Berry, and Hirst

1985). A number of utilities including BPA and Wisconsin Electric have found that loans can be difficult to administer. These utilities prefer grants, although some offer consumers a choice between grants or loans, so that those customers who prefer loans can participate in a program (Schick et al. 1990).

Multifamily Programs

Delivery of conservation services in multifamily buildings is hampered by the fact that incentives to invest in conservation improvements are often split between landlords, who own the building and benefit from capital improvements that add to property values, and tenants, who typically pay energy bills and thus benefit from bill reductions. Neither party receives enough of an incentive to proceed with investments on its own.

To deal with this problem, utilities have successfully employed three different program strategies. First, some utilities provide conservation measures to tenants at little or no cost. Many of the low-cost-measure programs discussed previously have been offered on this basis. Second, some utilities let landlords participate in commercial-sector programs (discussed below) for those energy uses subject to commercial rates—uses such as elevators and common-area lighting. Third, a few utilities market moderate-cost and comprehensive conservation services to landlords and provide sufficient financial incentive to encourage landlords to participate. An example of this approach is Seattle City Light's Multifamily Conservation Program, which provides technical and financial assistance to owners of electrically heated multifamily buildings. Financial incentives consist of ten-year, zero-interest loans, with a five-year deferred payment, and a 50% discount for first-year payoff. Funded measures include weatherization of individual units and improvements to lighting in public areas. After three years of program operation, 5% of eligible buildings have been served by the program (Shaffer et al. 1989). Preliminary estimates indicate that net savings from the program average about 12% of total building electricity use for shell and miscellaneous measures, and an additional 5% for lighting measures (Okumo 1990).

Low-Income Programs

Low-income customers are often targeted by utilities for special attention because (1) low-income customers are less likely to participate in many programs than higher-income customers, leading to the perverse result that low-income customers may subsidize higher-income participants; (2) low-income customers disproportionately contribute to utility arrearage costs, which can potentially be reduced if customer bills are lowered as a result of conservation actions; and (3) the utility or utility commission may decide that offering conservation services to low-income customers is good social policy.

Utility programs for low-income customers have taken two primary forms: (1) low-cost-measure programs aimed at a variety of end uses, and (2) comprehensive weatherization grant programs, usually based on the federal Weatherization Assistance Program (WAP). Low-cost-measure programs are offered either to low-income customers only or to all customers, with low-income customers receiving targeted marketing, often through community-based organizations, and special or free services. Grant programs may consist of supplemental grants for WAP participants, or grants may be targeted at moderate-income customers whose income slightly exceeds WAP guidelines (Morgan 1990). Experience with utility programs for low-income customers indicates that participation rates tend to increase when community-based marketing is employed using respected community organizations, when services are free (or heavily subsidized), and when measure installation is included among the services provided (Morgan and Katz 1984). Data on the cost-effectiveness of low-income programs are limited. Compared to programs aimed at the general population, low-income programs can be more, less, or equally cost-effective (NEES 1988a; Wirtshafter and Koved 1984).

New Construction Programs

While a home is being built, many conservation measures can be installed for only the incremental cost beyond standard construction practices. To retrofit these measures later is usually much more expensive and sometimes impossible. New construction conservation opportunities are often referred to as "lost-opportunity" resources, because if conservation measures are not installed at the time of new construction, many conservation opportunities are lost.

Residential new construction programs range from simple information programs to comprehensive programs that include education, technical training, and financial incentives. In most such programs the utility specifies performance or prescriptive standards for an energy-efficient home, certifies new homes that are in compliance with its standards, and promotes the advantages of certified homes to potential homebuyers through the media, cooperative advertising with builders, and other avenues.

In addition to these basic elements, some utilities provide rebates or rate reductions to builders or homebuyers; train builders on how to build and sell an energy-efficient home; and arrange, with local banks and secondary mortgage agencies, for higher loan limits for certified efficient homes (since reduced energy bills allow a homeowner to afford a slightly higher loan). Programs vary in their treatment of nonelectrically heated homes—some utilities certify and promote only electrically heated homes, while other utilities certify and promote all energy-efficient homes, although incentives are generally only paid for electrically heated homes.

Standards for what constitutes an energy-efficient home vary widely from utility to utility. Some programs target savings of only 10%, while others, like BPA's Super Good Cents and Canada's R-2000 program, target savings as high as 30–50% (Vine and Harris 1988). Programs that target high savings per home usually include financial incentives and extensive training programs. Some utilities primarily use their programs to promote heat pumps and exclude homes with electric resistance heat. Displacing electric resistance heat with heat pumps saves energy but may increase summer peak demand in homes that would not otherwise have air-conditioning. Displacing fossil fuels with heat pumps will increase electricity use. Only a few utilities have compared energy savings in homes built under their program with savings in nonparticipating new homes. Results thus far indicate that actual savings are sometimes higher and sometimes lower than engineering estimates (Schick et al. 1990; Gage and Niewald 1989). The accuracy of engineering estimates depends greatly on how accurately the baseline building—that is, one not participating in the program—is typified.

Participation rates in residential new construction programs vary widely. HERS participation rates range from 2% to 100% of new homes built (Vine and Harris 1988). Programs with high participation rates tend to target a small number of customers and to seek only modest energy savings in each home served. Two utilities in North Carolina have achieved participation rates above 50% with programs that offer 5–14% discounts on electric bills for certified efficient homes. In addition to the rate discount, the success of the programs has been attributed to the companies' efforts to educate consumers about the economic and thermal comfort benefits of additional insulation (Vine and Harris 1988). For programs that target large savings, participation rates are generally not as high. For example, the participation rate in BPA's program is 17–19%, although 20% of the utilities participating in BPA's program have achieved participation rates exceeding 40%, with one utility achieving a participation rate of 75–80% (Vine and Harris 1988; Schick et al. 1990). The more successful programs typically emphasize personal, one-on-one marketing and seek to enlist builders in the early stages of planning a project (Pat Zimmer, BPA, personal communication).

Program success requires (1) intervening early in the design and planning process; (2) including education, training, design assistance, and quality control activities; (3) obtaining the active support of builders, such as by involving builders in the program planning process and providing flexible program procedures that make it easier for builders to participate; and (4) actively marketing the program to consumers so that a demand is created for efficient homes (Vine and Harris 1988, Schick et al. 1990). In addition, programs introduced during a recessionary period are more apt to be successful because builders are receptive to novel ways to promote their buildings.

Data on program cost-effectiveness are not readily available. In a report on program costs for five residential new construction programs ranging from Austin's HERS program to BPA's comprehensive program, two programs emphasizing year-round conservation savings cost utilities $.02/kWh and $.031/kWh, but the assumptions behind these calculations were not made clear. For three programs emphasizing only summer load reductions, costs per kWh exceeded $.06, but costs per kW of peak load reduction were $200 or less (Schick et al. 1990).

A few utilities are using other approaches in addition to home certification and promotion programs to promote efficient new construction. The Town of Hull, Massachusetts, charges hookup fees that decline as the energy efficiency of the home increases, but due to a temporary moratorium on hookups for electric heat, the effectiveness of this approach has not been fully tested (Roger Jackson, Hull Municipal Lighting Plant, personal communication). BPA is unique in that it uses its Super Good Cents program as part of a long-term strategy to promote adoption of Model Conservation Standards—building code standards similar to the Super Good Cents certification requirements. Many municipalities and the state of Washington have adopted the Model Conservation Standards. Furthermore, several utilities have adopted the standards as a condition of providing electric service (Guy Hughes, BPA, personal communication).

Equipment Programs

Appliance Rebates

Many utilities have offered rebates for efficient residential refrigerators, freezers, air-conditioners, heat pumps, and water heaters. In a rebate program, the utility sets a threshold efficiency level and pays a rebate to customers who purchase an appliance whose efficiency exceeds the threshold. Participation rates, energy savings, and free riders vary widely from program to program depending on how efficient an appliance must be to qualify and on how effectively the utility markets the program. If efficiency thresholds are set too low, then a high proportion of available models qualify for rebates, resulting in high participation with a large number of free riders and low net energy savings per rebate due to the influence of free riders and to the fact that eligible appliances are only slightly more efficient than noneligible ones. These problems have plagued a number of rebate programs (McRae, George, and Koved 1988).

On the other hand, if eligibility levels are sufficiently high and marketing efforts strong, high participation levels, high energy savings, and a low proportion of free riders can be achieved. For example, New York State Electric and Gas (NYSEG) conducted a major refrigerator rebate experiment in 1985–1986. Only the 25% most efficient models offered by the industry at that time were eligible for rebates. Different marketing and rebate strategies were employed in different regions. Participation rates equalled 15% of refrigerator purchases in a control area, 35% in an information and advertising only area, 49% in a $35 rebate area, and 60% in a $50 rebate area. Dealer cooperation and promotion of efficient models was higher in the rebate areas and was considered critical to achieving high levels of participation (Kreitler and Davis 1987).

Recently, many utilities have discontinued refrigerator and freezer rebate programs because 1990 federal efficiency standards for appliances allow only efficient refrigerators to be sold in the U.S. These standards reduce the electricity use of a typical refrigerator by approximately 10% compared to a typical new unit produced before the standards took effect (Geller 1987). In 1993 more stringent refrigerator efficiency standards will go into effect, which will reduce refrigerator energy use by approximately 25% below 1990 levels (U.S. Department of Energy 1989).

However, even though federal standards are removing the least efficient models from the marketplace, several utilities (PG&E, BPA, NEES) still offer sophisticated programs to promote the most efficient appliances. For example, Pacific Gas and Electric (PG&E) offers rebates of $50 for refrigerators that exceed the 1990 federal efficiency standards by 10% and $100 for refrigerators that exceed the standards by 15%. PG&E estimates that over 30% of refrigerators sold in its territory in 1990 received rebates. Program costs were estimated to average $.031/kWh assuming a 20-year refrigerator life and 30% free riders (Gary Fernstrom, PG&E, personal communication). Unlike PG&E, BPA and NEES do not use rebates to promote high-efficiency refrigerators. For example, the BPA Blue Clue program awards an energy-efficient label to the 15% of refrigerators and freezers with the highest efficiencies in their classes. Qualifying models are promoted through consumer guides, media advertising, and stickers placed on showroom units (Anderson 1990). Estimates of energy savings due to the program are not available.

Another option being planned by several utilities is to offer high rebates for very high efficiency appliances that are technically feasible but not presently manufactured. This type of rebate offer—called a "golden carrot" because it provides a very lucrative carrot/incentive to manufacturers—is designed to provide a strong incentive for manufacturers to produce refrigerators that exceed the minimum efficiency standard by a substantial margin. For example, PG&E will provide rebates of $300 per refrigerator for units that exceed 1993 federal efficiency standards by at least 25% (PG&E 1990).

Four other appliance programs are worthy of mention because of the lessons they teach.

Wisconsin Electric Power Company (WEPCo) offers rebates for a variety of appliances including refrigerators and central air-conditioners. An analysis of electricity bills conducted to determine energy savings relative to a control group found that refrigerator savings were nearly identical with engineering estimates, while air-conditioner savings were much less than engineering estimates (Rogers 1989). Rogers postulates that purchasers of efficient air-conditioners may operate them more often than do customers who are unaware of or indifferent to how efficient their air-conditioners are. This tendency may be stronger in moderate climates where use of air-conditioning is considered discretionary.

A number of utilities offer programs to promote proper maintenance of air-conditioners and heat pumps. A pilot project conducted by the Salt River Project found that tune-ups performed using routine maintenance procedures resulted in average metered kWh savings of 7.7% of heat pump electricity use during the summer but had no noticeable impact on peak demand (Kuenzi and Wood 1987). Most of the savings came from a limited number of units that had performed poorly prior to the tune-up. Savings were less than the 10–20% expected, perhaps due to problems with many of the tune-ups (tuned air-conditioners operated below rated capacity and exhibited signs of refrigerant overcharge). These results indicate that successful tune-up programs require contractors trained in proper maintenance procedures. Also, use of screening procedures should be considered, so that only poorly functioning units receive extensive service.

PG&E and WEPCo, among others, have offered incentives for customers to retire old appliances in order to reduce the number of spare and backup appliances used in homes. WEPCo hires contractors to pick up old refrigerators, freezers, and room air-conditioners and gives customers a $50–$100 U.S. Savings Bond for participating. After three years, approximately 10–15% of eligible customers have participated, approximately the same rate as for the utility's rebate programs for new appliances. Customer surveys suggest that 50% of customers are free riders, who would have disposed of their old appliances on their own (Schick et al. 1990). Typically the utilities donate to charity those units whose efficiency exceeds the average for used appliances, and they destroy the less efficient units.

Both BPA and NEES have run programs to promote high-efficiency water heaters (heat pump and solar water heaters in BPA's case, high-efficiency conventional water heaters in NEES's case). In both cases participation levels were less than 10% of water heater sales because plumbers appeared uninterested in changing work procedures, despite very high rebates, including rebates paid directly to plumbers and wholesalers; extensive marketing; and streamlined application procedures. However, manufacturers were much more responsive than plumbers to the NEES program. When the program began, only 40 models met its efficiency thresholds. After meetings with NEES, however, manufacturers introduced 33 new models during the first six months of the program (Major and Cody 1987; NEES 1988b; Bernie Mittelstaedt, DEC International, and Pat Stell, NEES, personal communications).

Lighting Programs

Most residential lighting programs emphasize compact fluorescent lamps because these products can achieve savings of up to 75% without reducing light output and because they last approximately ten times longer than standard incandescent lightbulbs, thereby providing the utility with long-term savings. A few programs also promote energy-saving incandescent lamps, although these products typically save only 10% or so. Compact fluorescents are difficult to promote because single lightbulbs cost $8–$20, approximately 20 times the cost of a standard incandescent bulb. Also, compact fluorescent lamps are generally not available in supermarkets and other retail outlets.

Despite these difficulties, a number of highly innovative programs have achieved considerable success in promoting compact fluorescent lamps through techniques such as mail-order sales, charity sales, coupons, and leasing. For example, WEPCo achieved a 7% participation rate in one year with a mail-order program. Bulbs were purchased in bulk by the utility and sold to customers for an average of $2.50 apiece. CMP achieved a participation rate in excess of 20% by promoting efficient lamps through the Lions Club annual lightbulb sale. Bulbs were purchased in bulk by the utility and sold to consumers at a discount of 72–84% off the normal retail cost. Available supplies were sold out in two weeks (Schick et al. 1990). Many Scandinavian and Dutch utilities have recently begun to promote compact fluorescent lamps to their retail customers. For example, Stockholm Energi ran two five-week campaigns that combined extensive publicity, coupons good for a $7.50 discount on any compact fluorescent lamp, and voluntary wholesaler discounts of 30%. The participation rate for the two campaigns was approximately 20% (Evan Mills, Lund University, personal communication). The Taunton Municipal Lighting Plant in Massachusetts leases compact fluorescent bulbs to customers at a cost of $.20/month included on the electric bill and guarantees that customers will save more than $50 in energy costs over the life of the bulb. Five percent of residential customers have participated, with each participant leasing an average of four bulbs (Joe Desmond, Taunton Municipal Lighting Plant, personal communication). The Taunton program has reportedly cost the utility $.025/kWh (Krause, Vine, and Gandhi 1989). Rough calculations for some of the other programs described above indicate similar costs.

In addition to the program approaches discussed above, giveaway programs have been successfully employed to promote compact fluorescent lamps, as discussed previously in the section on low-cost measures.

Commercial and Industrial Programs

The commercial and industrial sectors account for approximately 28% and 34% respectively of U.S. electricity sales (EIA 1989), though these proportions vary widely among utility service territories. In the C&I sectors, a limited number of medium- and large-sized customers use most of the energy, and a large collection of small customers uses the remainder. For example, Niagara Mohawk reports that 91% of C&I customers use less than 100,000 kWh/year and account for only 18% of C&I electricity sales (Flaim 1990). Nationwide, approximately 37% of commercial electricity use goes to lighting, 18% to ventilation, 15% to air-conditioning, 13% to space heating, and 17% to water heating and other miscellaneous uses (BNL 1987).

This section discusses C&I programs targeted at existing buildings (including audit, information, rebate, loan, performance contracting, and comprehensive direct installation programs). It also describes programs developed for new buildings and those designed specifically for industrial customers (although many of the programs for buildings are also relevant to industrial customers). Information in this section draws extensively from *Lessons Learned,* a recent study of more than 200 C&I programs (Nadel 1990a). Unless otherwise noted, all data are from this report.

Existing Building Programs

Audit Programs

C&I audit programs have been offered by many utilities. The typical C&I audit program combines a nonengineering audit (detailed engineering assessments are not made) with limited marketing and postaudit follow-up efforts. Participation rates are typically 1% per year (cumulative participation less than 10%), but several programs have achieved participation rates in excess of 60%, including programs operated by the Sacramento Municipal Utility District (SMUD), Southern California Edison, Con Edison, and Massachusetts Electric. These programs all combined free energy audits with extensive personal marketing efforts.

The importance of free audits and personal marketing is illustrated by an experimental study conducted by NYSEG (Xenergy 1990). In this study some customers were offered free audits, some charged a fee; some customers were marketed by phone and mail, some via personal contacts by utility representatives. Free-audit groups averaged 50% participation, fee groups averaged less than 20%. Personal marketing groups averaged 37% participation, phone and mail groups only 9%. Combining free audits and personal marketing achieved a participation rate of 65%.

Several studies, some based on engineering estimates and some on billing studies, have found that customers who receive audits reduce their kWh use by an average of 2–8%. Results based on billing analysis generally correspond to those based on engineering estimates, although no published study compares engineering estimates with billing analysis for the same program. The highest savings were achieved by programs that included periodic postaudit visits to participating customers to determine which measures had been installed and to recommend additional measures. Several of the programs with above-average savings also included financial incentives to help pay installation costs.

C&I audit programs have low costs per kWh saved, generally less than $.02/kWh, an estimate assuming a five-year average measure life but not adjusted to eliminate free riders. To keep programs for small C&I customers cost-effective, many utilities use streamlined audits emphasizing low- and no-cost measures. For large customers, experiences by several utilities indicate that detailed engineering audits may not be worth the expense unless customers are seriously interested in implementing audit recommendations (Southern California Edison 1983; Cambridge Systematics 1988).

Other Information Programs

Besides audits, information programs can range from simple educational brochures, distributed by mail, to sophisticated marketing programs. Utilities have run hundreds of such programs but rarely have compiled or published results. Limited data available indicate that information programs can have a positive but limited impact in terms of participation rates and energy savings.

For example, Orange and Rockland Utilities in New York/New Jersey mailed an informational brochure on lighting efficiency improvements to 18,000 customers, and fewer than 1% responded by sending in a tear-out card to request additional program information (Orange and Rockland 1988). Similarly, Niagara Mohawk mailed an informational brochure on energy-saving fluorescent lamps to a targeted group of lighting decision makers at customer facilities. In a survey conducted at the end of the six-month experiment, 3% of these customers reported they had switched to high-efficiency fluorescent lamps in the last six months, and 5.6% of customers who received the same brochure combined with a rebate offer reported switching (Clinton and Goett 1989).

Rebate Programs

Rebates are the most common type of financial incentive programs offered by electric utilities. In particular, lighting rebate programs and multiple end-use rebate programs—programs that provide rebates for measures affecting several different end uses—are commonly offered by utilities. Air-conditioning and motor rebate programs are also common. Most rebate programs pay rebates equal to 20–50% of the cost of a measure.

Most rebate programs have achieved cumulative participation rates of less than 4% of eligible customers. The most successful rebate programs, such as the multiple end-use rebate programs offered by PG&E and the city of Palo Alto, California, have served approximately 5–10% of C&I customers, including approximately 20–25% of large customers (customers with peak demand greater than 100–500 kW). These high-participation programs feature simple application procedures, catchy marketing materials, active involvement of equipment dealers and other trade allies, free energy audits to help customers identify conservation measures, and extensive personal marketing with an emphasis on developing a personal relationship with customers, especially large customers.

The most successful rebate programs have been effective at promoting basic lighting improvements and improved-efficiency HVAC units. Most rebate programs currently in operation have devoted limited or no attention to promoting advanced technologies or *system* improvements, which involve the interaction of multiple pieces of equipment.

High-efficiency motors have proven more difficult than lighting and HVAC equipment to promote through rebate programs. Most motor rebate programs have affected less than 5% of motor purchases. However, B.C. Hydro has affected an estimated 30% of motor purchases and hopes to reach 50% with a program that combines rebates with a multifaceted education program. The latter includes an educational booklet for customers, computer software for dealers and large customers to use to estimate energy savings, a list of all dealers in the province supplying efficient motors, and a database that classifies by type and efficiency level all motors available in the province. To market the program, utility field representatives personally contact large customers, consulting engineers, and motor dealers. As a result, many motor dealers are actively promoting the program (Kristin Schwartz and Jack Habart, B.C. Hydro, personal communication).

Data indicate that free rider shares in rebate programs vary widely, from less than 5% to nearly 90%, although most of these estimates are based on customer self-reports, which are subject to errors resulting from trying to please the interviewer, answering with the socially acceptable answer, and failing to recall purchase decisions made many months earlier. As might be expected, free ridership is higher for products already in demand (such as reduced wattage fluorescent lamps) and lower for products to which consumers have been less attracted by market forces alone (products such as compact fluorescent lamps or electronic ballasts). Thus, as with residential appliance rebates, free rider proportions can be reduced to a large extent by selecting appropriate products and efficiency levels. Of course, as efficiency thresholds increase, participation rates can be expected to decline, at least initially.

Another technique several utilities employ to reduce free riders is to pay incentives only for measures that exceed a specified payback period, typically one year. This approach screens out highly cost-effective measures that customers are likely to implement on their own. On the other hand, this approach is more difficult for customers to understand and for utilities to administer. Due to these limitations, this approach is primarily used in programs targeted at large C&I customers who have the necessary staff and skills to work with a more complex program (Weedall and Gordon 1990).

At times, high free rider levels can be acceptable. For example, NEES ran a rebate program to promote energy-saving fluorescent lamps, a product that had a market share of approximately 45% prior to the beginning of the program. In the first year of the program, 60–80% of the participants were free riders. Still, the program cost the utility $.017/kWh and had a benefit-cost ratio of 2.0 from the utility perspective (Nadel 1988).

The effect of rebate level on participation rates has been widely discussed. Research suggests that many factors affect program participation and that often the level of financial incentives is not the most important factor. When program features are held constant and only the level of incentives is varied, however, participation rates tend to increase as financial incentives increase (Berry 1990). Other work also shows that incentives and participation levels may not be simply and directly related: on the one hand, adding incentives to a no-incentive program tends to increase participation rates, and incentives greater than 50% of measure cost can increase participation rates above levels achieved with low incentives; on the other hand, the difference between low and moderate incentives (approximately 15% and 30% of measure cost respectively) may have little effect on participation rate (Nadel 1990a).

Most rebate programs pay customers, although a few programs have paid dealers under the premise that if dealers received the rebates, they would have greater incentive to promote rebate-eligible products. Results of dealer rebates have been mixed. For example, both Southern California Edison and Eastern Edison received significant response from air-conditioning dealers who were offered rebates or points redeemable for gifts for each efficient product they sold (Bruce Mayo, Southern California Edison, and Carol White, Eastern Edison in Massachusetts, personal communications). On the other hand, Southern California Edison, Nevada Power, and Northern States Power (NSP) in Minnesota all received limited responses when incentives were offered to motor dealers (Bruce Mayo, Southern California Edison, Bob Tyre, Nevada Power, and Randy Gunn, NSP, personal communications). However, the limited response of motor dealers could have been due to the low level of rebates paid ($10/motor, $.50/horsepower, and points redeemable toward gifts). NEES obtained mixed results from a program that paid rebates to lighting dealers. Some dealers actively promoted the program by hiring extra marketing staff and passing rebates on to customers. Other dealers did little promotion and kept the rebate themselves. Overall, participation rates were somewhat higher than a pilot customer rebate program NEES had previously run (Nadel 1988; John Eastman, NEES, personal communication).

Few rebate programs have collected data on energy savings as a percent of the preprogram electricity use of participating customers, although a few studies provide limited indications of the savings that can be achieved. Participants who received audits and rebates from Southern California Edison's Hardware Rebate program netted average savings of 7% above savings achieved by customers who received only the audit (actual measured savings were very close to prior engineering estimates) (Train and Strebel 1986). BPA's Commercial Incentive rebate program for small- and medium-sized C&I customers netted average savings of 6% (Dagang 1990). Measured savings averaged only 36% of prior engineering estimates. Reasons for the discrepancy include overly optimistic savings estimates; removal of measures by some program participants; and interactions between measures and between measures and HVAC systems—interactions not factored into the engineering estimates (Ken Keating, BPA, personal communication).

Rebate programs are typically inexpensive for the utility, with median costs for most rebate programs of approximately $.01/kWh, assuming a ten-year average measure life. These figures are generally based on engineering estimates and are not adjusted for the effect of free riders. Even programs that pay high rebates generally cost the utility less than $.02–.03/kWh.

Loan Programs

C&I loan programs have only been offered by a few utilities. Side-by-side comparisons with rebate programs offered by the same utilities show that most customers prefer rebates, although loans are useful for the minority of customers who lack cash to finance energy-saving investments. For example, both Wisconsin Electric and Puget Sound Power & Light offer customers a choice between a zero-interest loan or a rebate approximately equivalent to the interest subsidy on the loan. In both programs, over 90% of the participating customers have chosen rebates instead of loans (Peggy Clippert, Wisconsin Electric, and Syd France, Puget Power, personal communications).

Performance Contracting Programs

Performance contracting programs generally rely on energy service companies (ESCOs) to provide services to customers. ESCOs receive payments from the utility for each kWh or kW they save. Participation rates in performance contracting programs have ranged from less than 1% to as high as 15% (for a pilot program offered to a limited group of customers). Programs with high participation rates and savings, such as those offered by Boston Edison, NEES, and Commonwealth Electric, generally provide ESCOs with payments that approach the utility's avoided costs. For example, Boston Edison's ENCORE program paid an average of $.033/kWh in 1988 (payments vary with measure life). With these incentives, ESCOs can provide many measures at no cost to the customer and still make a profit (Steven Murphy, Boston Edison, personal communication).

Left to their own devices, most ESCOs will choose to concentrate on the largest customers and the most lucrative energy-saving measures, particularly lighting and cogeneration. Both Boston Edison and NEES found that ESCOs are primarily interested in customers with peak demands of 500 kW or more. In both programs, ESCOs tended to emphasize lighting measures. NEES found a strong ESCO interest in cogeneration systems as well (NEES 1988a and 1988b; Steven Murphy, Boston Edison, personal communication).

Limited side-by-side comparisons indicate that other program approaches will achieve greater participation than ESCO-based programs. For example, NEES operated both a performance contracting program and a customer rebate program for several years. Participation rates were approximately 40% greater for the rebate program, and kW savings were approximately 165% greater for the rebate program, despite the fact that the performance contracting program paid higher incentives (45% higher on average), had been operating six months longer, served larger customers (the performance contracting program targeted customers with billing demand greater than 500 kW, while the rebate program had a 100-kW threshold), and served twice as much total customer demand (Hicks 1989).

Furthermore, performance contracting programs are generally more expensive to the utility than utility-operated programs promoting the same measures, because the utility must directly or indirectly cover ESCO overhead and profit. For example, the median utility cost per kW and kWh saved was approximately three to four times higher for eight performance contracting programs examined by Nadel (1990a) than for 16 multiple end-use rebate programs.

Many utilities—including BPA, NEES, and Wisconsin Power and Light—have either phased out performance contracting programs or complemented them with other types of programs. Still, performance contracting programs can be useful both for customers who do not have the time, money, or expertise to implement energy-saving measures on their own and for utilities who prefer having outside contractors deliver energy services to administering programs themselves (although work with ESCOs provides its own hassles).

Comprehensive Programs

Comprehensive programs generally provide one-stop shopping to the customer. Services may include audits; arranging for measure installation; financing assistance (loans or grants); and sometimes operations, maintenance, and other follow-up services. These programs are designed for customers who lack the time, money, or expertise to identify and implement conservation projects on their own.

Comprehensive programs can achieve higher participation rates than other program approaches used to date, although many of the most successful programs have only been operated on a highly targeted basis. For example, Southern California Edison, the Snohomish Public Utility District in Washington, and Boston Edison have all achieved participation rates of approximately 70% with programs targeted respectively at 855 large commercial customers, 35 nonprofit customers, and ten large commercial customers (Nadel 1990a).

Sacramento Municipal Utility District and Massachusetts Electric have achieved participation rates of 55% and 34% of targeted customers (small C&I customers excluding minimal-use customers such as phone booths and billboards) with programs that provide free audits and installation of energy-saving lighting improvements at no cost to the customer. Due to a need to assign resources to other programs, both utilities ended their small C&I lighting programs before all eligible customers could be reached (Neos Corp. 1989; Nadel 1988; Kathy Itow, SMUD, personal communication; John Oinonen and Don Robinson, NEES, personal communications). A billing analysis of the Massachusetts Electric program found that participants reduced their electricity use by 9% on average. Actual savings for the program were nearly identical to engineering estimates, although small customers saved less than the engineering estimates, and larger customers saved more than the engineering estimates (Nadel and Ticknor 1989).

Several utilities in the Pacific Northwest have offered comprehensive programs to all C&I customers. These programs typically have participation rates of approximately 1%/yr. In these cases, program participation has been limited by budget considerations (Steve Poole, Seattle City Light, and Syd France, Puget Power, personal communications).

Comprehensive programs can also achieve higher energy savings than rebate programs. Savings estimates for five different multiple end-use comprehensive programs ranged from 10% to 23% (all but one of these estimates were based on engineering data). BPA's Commercial Incentives Pilot Program averaged actual net savings of 26% among large facilities that received comprehensive services (actual savings were 9% greater than engineering estimates) (Dagang 1990). In general, programs with the highest savings include special efforts to implement all cost-effective measures; programs without these special efforts generally achieved savings of 10–15% (Nadel 1990a).

The high participation and energy savings achievable by comprehensive programs come at a price: comprehensive programs typically cost utilities $0.02–0.04 per kWh saved (assuming a ten-year measure life), a price below the long-term avoided cost of many utilities but above the cost to a utility of a typical rebate program. However, when customer costs are included in the analysis, rebate program costs per kW and kWh begin to approach comprehensive program costs because customers pay a higher share of measure costs in a typical rebate program than in a typical comprehensive program.

New Construction Programs

New commercial construction can be a major source of load growth for electric utilities. For example, NEES has estimated that 75–80% of commercial sector load growth over the 1987–1997 period will be due to new buildings as opposed to additional energy use in existing buildings (NEES and CLF 1989). As with residential new construction, the rationale for offering a C&I new construction program is that it is generally easier and less expensive to incorporate energy-saving measures at the time of building construction than to retrofit a building after it is completed. Costs per kW saved can be as much as 80% lower when measures are incorporated into new construction instead of being retrofit (ibid.).

C&I new construction programs fall into three main categories:

technical assistance programs, rebate programs, and comprehensive programs.

New construction technical assistance programs assist building designers to improve the energy efficiency of their designs. Typical services include workbooks, educational seminars, and free computer simulations of how much energy a building will use under different design scenarios. Some informational programs also include awards to recognize the designers and developers of exemplary buildings. Examples of technical assistance programs include TVA's C&I New Construction Program and BPA's Energy Smart program. Information on participation rates and percentage savings achieved is not available. Operators of many technical assistance programs report considerable success in convincing developers to adopt energy-saving recommendations, although they believe that measure adoption rates could be improved if financial incentives were offered (Jim West, TVA, and Syd France, Puget Power, personal communication). The TVA program cost the utility $.034/kWh, assuming a 20-year average measure life (Nadel 1990a).

New construction rebate programs provide rebates for incorporation of specific measures into new buildings. Common measures include high-efficiency lighting fixtures, motors, and cooling systems. Some programs provide incentives for measures proposed by the customer or building designer. Typically, savings from these measures are evaluated relative to local building code requirements or to prevailing local construction practices. Some rebate programs also include technical assistance provided by utility staff or private consulting firms on retainer. Examples of rebate programs include Wisconsin Electric's Smart Money New Construction program and Southern California Edison's Energy Excellence program. Data on participation rates and percentage savings are not available. Utility costs average approximately $.01/kWh saved, including free riders. Wisconsin Electric estimates that free riders account for approximately 30% of the savings in the utility's C&I new construction rebate program (Wisconsin Electric 1989).

New construction comprehensive programs combine technical assistance and rebate program features. These programs generally include training and technical assistance services, free computer simulations, construction incentives, incentives for additional design time undertaken by the project design team, and postconstruction building commissioning and monitoring services. Most of these programs pay the full incremental cost of efficiency measures not normally included in standard construction practice. Examples of comprehensive programs include BPA's Energy Edge program,

Northeast Utilities' Energy Conscious Construction program, and NEES's Design 2000 program. Full-scale comprehensive programs have been in operation for only a short time, so participation rates are not available. BPA's pilot program Energy Edge reduced the energy use of participating office buildings by 21% compared to prevailing construction practice. Savings were approximately 25% lower than prior engineering estimates due primarily to problems with quality control and commissioning of installed measures (Harris and Diamond 1989). Design and construction costs totaled approximately $.027/kWh saved, based on actual savings and assuming a 20-year measure life but excluding utility administrative costs (Anderson and Benner 1988).

Seventeen successful C&I new construction programs examined by Nadel (1990a) were found to feature most or all of the following elements, regardless of the type of program: (1) market research to identify C&I construction practices and trends, so that program requirements and marketing efforts can be properly targeted; (2) personal marketing to architects, engineers, and developers; (3) efforts to enroll participants early in the design process, before design and specification decisions are made; (4) extensive training of and technical assistance to the project design team; (5) publicity for the designers and developers of successful energy-efficient projects; (6) technical assistance and monitoring through the construction and project start-up stages; and (7) financial incentives to help pay the incremental cost difference between standard- and high-efficiency equipment and designs.

Industrial Programs

While most utilities design programs to serve both commercial and industrial customers, many utilities are finding that combined C&I programs primarily promote basic lighting, HVAC, and motor improvements, and that only limited industrial process improvements are being implemented. For example, 37% of Wisconsin Electric's kWh sales are to industrial customers (EIA 1989), but only 10% of kWh savings achieved by their C&I Smart Money program are due to industrial process improvements, including high-efficiency motors (Wisconsin Electric 1989).

Barriers to industrial customer participation in conservation programs include (1) program marketing materials that emphasize commercial conservation measures and thereby give the impression that the program has little to offer industrial customers; (2) concerns about shutting down process lines in order to install new equipment (the value of a single day of production can equal an entire year of energy savings); and (3) the fact that most industrial process improvements require engineering analysis and supervision, but many plants and most utilities lack staff with the necessary skills or time to undertake such projects.

To deal with these problems, a few utilities have begun to offer programs targeted toward industrial process improvements. For example, BPA, CMP, and Northeast Utilities (NU) offer programs where customers propose projects to the utility, and the utility pays an incentive per kWh saved. In BPA's program, customers first submit short project abstracts, and only if the abstract looks promising is the customer asked to prepare a detailed proposal. All three utilities only fund measures whose payback exceeds a utility-set threshold. The purpose of the threshold is to encourage customers to implement measures they would not be likely to pursue on their own. A process evaluation of the BPA program recommended increased field visits and technical assistance to encourage participation, a rapid review process for proposals submitted by customers, and a simple contract (Peters 1989).

BPA offers two other unique programs targeted to industrial customers. The Design Wise program provides engineering reviews of new construction and expansion plans in order to make recommendations on ways to improve process lines before they are built. The Aluminum Smelter Conservation/Modernization program provides incentives for large aluminum smelters to implement conservation measures. Eligible customers were heavily involved in the program design process. As a result of this and other factors, nine out of ten eligible customers elected to participate in the Aluminum Smelter program (Tom von Muller, BPA, personal communication).

NEES has recently announced plans to offer several new services to industrial customers in order to encourage participation in existing C&I programs. Services include technical assistance provided by experts in particular process industries and a lending library of energy test equipment to allow customers to more accurately estimate energy losses and potential energy savings (NEES and CLF 1989).

Programs targeted specifically to industrial customers are still in their infancy. The 17 industrial programs examined by Nadel (1990a) were found generally to have low participation rates (less than 10%, except for the Aluminum Smelter program) and low utility costs per kWh (median costs for ten industrial programs were $.008/kWh, assuming a ten-year average measure life and including free riders). Further work is needed to improve participation rates, but the low cost of the typical program indicates that there is considerable room to maneuver and still keep programs within reasonable cost constraints.

Bidding Programs

In the past few years there has been considerable interest in bidding programs where utilities request proposals from outside parties to supply demand-side or supply-side resources or both. Successful bidders are selected on the basis of price and other factors. The purpose of bidding programs is to let the market determine the price of new resources and the proper mix of program efforts, including the mix between demand- and supply-side resources and the mix of utility and nonutility programs. In some bidding programs, bids are limited to specific sectors (such as C&I), while in other programs, bids for any sector can be submitted.

In developing bid programs, a number of issues must be addressed including resources and sectors to be solicited, bid evaluation criteria, bid ceiling price, and how savings will be measured (Goldman and Hirst 1989).

Experience with demand-side bidding programs is limited, although bidding programs are proliferating rapidly. Initial experiences indicate that bids are primarily for large C&I projects rather than for residential and small C&I projects. Most demand-side bids have been submitted by ESCOs, although some bids have been submitted by large C&I customers (Estey 1989). Indications thus far are that these programs can achieve significant energy savings. For example, by the end of 1989, CMP had signed contracts totaling 1.5% of its peak demand through its Power Partners and Efficiency Buyback programs (Jonathan Linn, Central Maine Power, personal communication). Bidding programs cost less than utility avoided costs because bid prices are capped at avoided costs, although there is a tendency for bids to approach utility avoided costs (Estey 1989).

While further experience is needed before definitive conclusions can be drawn, initial experiences indicate that demand-side bidding programs may have a limited role to play in a utility's overall demand-side management strategy due to the limited development of the energy services industry in the United States, to high transaction costs in certain sectors such as the residential and small commercial sectors, and to the inappropriateness of bidding mechanisms for various types of programs (such as design assistance for new construction and other informational programs) (Goldman and Hirst 1989). Thus, initial indications are that bidding programs will play a role in providing demand-side resources, particularly for utilities that are unable or unwilling to operate their own programs, but in most cases, bidding programs will represent only one part of a comprehensive package of demand-side management programs.

Discussion

From the preceding discussion of the lessons taught by residential, commercial, and industrial conservation programs, a number of common threads emerge. Some of these key points, and their implications, are summarized below:

1. Energy audits designed to help customers identify conservation opportunities generally result in only limited energy savings. Implementation of audit recommendations is generally increased by the provision of postaudit follow-up services (periodic follow-up calls, arranging installations) and financial incentives. Audit complexity should be based on customers' annual electricity use and on the measures that customers are likely to implement given the follow-up services available.
2. Different program approaches fill different niches. Rebate programs can be successfully used to promote efficient equipment at a moderate cost to the utility. However, rebate programs generally only reach a minority of customers and have not been very effective at promoting improvements involving the complex interactions of multiple pieces of equipment. Loan and performance contracting programs can be useful for the minority of customers who lack capital to finance conservation improvements. Performance contracting and bidding programs are also useful for utilities that do not want to operate programs on their own. However, these program approaches require extensive administration and oversight by the utility, and are apt to be costlier than utility-managed programs. Comprehensive direct installation programs can achieve higher participation rates and energy savings than most other program approaches, but generally at a higher cost to the utility (though not to society) than information, rebate, and loan programs. This approach may be particularly suitable for utilities serving hard-to-reach customers (like low-income residents and small C&I customers) or for utilities with capacity needs in the short and medium term.
3. Marketing strategies and technical support services have a large impact on program participation and savings. One-on-one and intensive community-based marketing strategies can be particularly effective. Equipment dealers, contractors, and design

professionals can be important allies in promoting programs. In designing programs, it is important to keep customers' needs in mind and to make sure that marketing materials and program participation procedures are easy for customers to understand.

4. All other things being equal, financial incentives tend to increase program participation and savings. High incentives appear to promote greater participation than moderate incentives, but the impact of low and moderate incentives may be indistinguishable.
5. An important target for utility programs is lost opportunity resources—situations such as new construction where there is a one-time opportunity to purchase conservation savings at a low cost. Other lost opportunity situations include times when major equipment is replaced and when buildings or process lines are remodeled. Lost opportunity programs should be considered by all utilities, even utilities with large capacity surpluses.
6. The number of free riders in a program depends in large part on program design, including which measures are promoted; on where efficiency levels are set; and on how incentives are structured. When measures with high current market shares and/or rapid payback periods are promoted, free riders tend to be high; with products with low market shares or less rapid payback periods, free riders tend to be low. In some cases, even programs with high levels of free riders are cost-effective to the sponsoring utility.
7. Measured savings from residential retrofit programs are often less than engineering estimates, especially in homes with extensive use of wood or other secondary heat sources. Available data on commercial and new construction programs are limited and show that savings are sometimes higher and sometimes lower than engineering estimates. However, even when actual savings are less than estimates, the program may still be cost-effective to the sponsoring utility or society or both.
8. There are many gaps in our knowledge about how to design and operate successful utility conservation programs. While many creative program approaches have been tried and evaluated, there is a need for new approaches to increase the number of customers reached and measures implemented. Also, because many existing programs are not fully documented and evaluated, important information is unavailable. These issues are discussed more extensively in the last section of this chapter.
9. When one examines successful programs the names of a few

utilities—BPA, CMP, NEES, NU, Puget Power, Seattle City Light, and Wisconsin Electric— come up again and again. The commitment of these utilities to making conservation programs work is shown by their willingness to thoroughly research and design new programs, modify existing programs, and conduct regular evaluations to learn the lessons each program teaches. As is discussed in the next section, utilities with this level of commitment are often the utilities whose programs have achieved the greatest energy and demand savings. Thus, it appears that the presence of a high level of utility commitment to conservation programs is a major determinant of program success.

Putting the Experience to Date in Context

A recent analysis published by the Electric Power Research Institute (EPRI) estimates that in 1990 DSM programs will save a total of 32,955 GWh, will reduce summer peak demand by 19.3 GW, and will reduce winter peak demand by 14.8 GW (Faruqui, Seiden, and Braithwait 1990). These savings represent 1.1% of projected 1990 GWh sales, 3.6% of projected summer peak demand, and 3.0% of projected winter peak demand. This report goes on to estimate that in the year 2000, the most likely impact of DSM programs will total 3.0% of GWh sales in that year, and 6.5% of summer and 5.3% of winter peak demand.

While the average utility is reducing electricity use and demand by 1–4%, and is projected to increase these savings to 3–7% in the next ten years, some utilities are doing considerably better. Seven utilities with particularly long-term, large, and aggressive conservation and load management (C&LM) programs (Austin, CMP, NEES, PG&E, Seattle City Light, Southern California Edison, and Wisconsin Electric) have achieved energy savings ranging from 1.4% to 13.7% of 1987 kWh sales and from 2.5% to 8.5% of 1987 peak demand (Geller and Nadel 1989). These figures are generally based on engineering estimates and in some cases include free riders. Within this group, Southern California Edison and PG&E have achieved cumulative energy and demand savings of 6–14% over eight years. Other utilities that have more recently begun large-scale DSM efforts are reducing energy and demand by 0.6–1.4% annually.

The savings achieved by the most aggressive utilities, though substantial, are significantly less than the technical potential for conservation savings. A recent study on the technical potential for C&LM savings in New York state estimated that if all C&LM

measures cost-effective to consumers (assuming a 6% real discount rate) were implemented, statewide electricity use would be reduced by 35% (Miller, Eto, and Geller 1989). Another report, prepared for EPRI, estimated a technical conservation potential (ignoring economic considerations and many implementation constraints) of 24–44% using technologies that are readily available today (Faruqui et al. 1990). Additional studies have estimated a technical conservation potential ranging from 20% to 75% (Faruqui et al. 1990; Geller and Nadel 1989). Thus, for the average utility, the kWh savings achieved in 1990 probably represent less than 5% of the technical C&LM potential (1.1%/34%), and even for the most successful utilities, conservation savings achieved represent on the order of one-third (10%/34%) of the available resource.

While achieving all of the technically available conservation savings is unlikely due to a variety of economic and implementation constraints, the estimated potential indicates that even the most successful utilities can probably tap additional conservation resources, and the savings achieved by the most successful utilities indicate that large savings are available to the average utility.

Future Directions—Issues for the Next Decade

Over the past decade, utility conservation programs have made substantial progress. Significant savings have been achieved, and a large amount of information on the results of different program efforts has accumulated. This experience teaches us many important lessons about ways to structure and promote programs in order to achieve substantial energy and dollar savings. These lessons will prove very useful during the 1990s, as many utilities expand their conservation efforts. However, much remains to be learned if even half the technical potential for conservation improvements is to be achieved. Important issues to be addressed in the 1990s include the following:

Evaluation

As utilities come to increasingly rely on demand-side resources to meet energy and capacity needs, carefully documenting program savings and cost-effectiveness will become increasingly important. Hirst (1990) argues that utilities should collect and report information on demand-side resources comparable in content and detail to the data on supply-side resources that utilities annually report to the Energy Information Administration (EIA) and the Federal Energy Regulatory Commission (FERC). In particular, information on participation rates, free rider shares, savings (both kWh and coincident kW), and costs should be collected and reported for all programs. Whenever possible, savings and free rider data should be based on a statistical analysis of participants' and nonparticipants' measured electricity use. Standard definitions for key terms such as participation rate, indirect costs, and free rider need to be established, so data for different programs can be compared.

Improved Program Approaches

Available evaluation data, though limited, indicate that most programs are serving less than 10% of eligible customers and achieving savings among participating customers of less than 10%. Research and experimentation are needed on ways to improve participation rates. Comprehensive programs have shown promise in this regard, primarily in pilot and limited-scale applications. Comprehensive approaches need to be implemented in large-scale efforts, and new, creative approaches tried. For example, utilities should develop long-term strategies to bring new technologies and design strategies into widespread application. An example of such a strategy is BPA's work in residential new construction, which combines education and incentive programs with work on building codes, in an effort to foster a major change in prevailing construction practices over a period of about ten years. Improved approaches are also needed to promote industrial measures and system improvements involving the interaction of different types of equipment. These measures are generally not being successfully promoted by current programs. Also, improved program approaches are needed to capture the lost opportunity resources available at the time of new building construction, building remodeling, and major equipment replacement.

Interactions with Other Utilities

This chapter has discussed only programs operated by electric utilities to reduce electricity use. Gas and water utilities are also offering conservation programs, often to the same customers that are served by electric utility programs. In some situations it may be advantageous for utilities to work together on a single program, with costs allocated according to the resources conserved. For example, United Illuminating is working with local water companies on a program that installs low-cost electricity-saving and water-saving measures in the same home visit (Blair Hamilton, Vermont Energy Investment Corp., personal communication). Several utilities and utility commissions, such as the Vermont and Wisconsin commissions, are investigating fuel-switching as a conservation option. At times, energy and money can be saved by encouraging customers to use one fuel instead of another for a particular end use. Cost-effective fuel-switching options vary as a function of climate and local energy rates, but may include switching from fossil fuels to electricity for certain industrial processes (Faruqui et al. 1990) and switching from electricity to fossil fuels for space and water heating (Chernick, Goodman, and Espenhorst 1989).

Incentives for Utilities

There is growing recognition that utilities lose money when they effectively promote electricity conservation: cutting electricity use reduces revenues and profits in the short run (Moskovitz 1989). A few states have started to take steps to overcome this barrier by offering utilities incentives such as bonus payments per kWh saved for pursuing C&LM programs incorporated in least-cost plans. Policies that make the least-cost strategy also the most-profit strategy for the utility could go a long way toward convincing utilities to vigorously promote and finance C&LM programs. These policies should be encouraged and then thoroughly evaluated to determine whether utility incentives promote improved C&LM programs and which incentives best achieve particular objectives.

Environmental Initiatives

In recent years concern about air pollution and global warming has increased. Amendments to the Clean Air Act will allow utilities to take credit for conservation programs when demonstrating compliance with sulphur oxide and nitrogen oxide emissions requirements. Future legislation may regulate carbon dioxide emissions in the same manner. These and similar environmental initiatives provide utilities with additional justification to undertake conservation programs, but these initiatives bring an added responsibility to accurately measure program impacts, including pollutant reductions, so that proper emission credits can be determined.

Relationship between Utility Programs and Codes and Standards

Even improved utility programs cannot achieve all of the cost-effective savings that are technically achievable. Some customers will always choose not to participate in a program. Building codes, equipment efficiency standards, and other similar policies can achieve additional savings beyond those achieved by utility programs. For example, the California Energy Commission estimates that in 1983, C&LM measures and practices reduced peak demand requirements in the state by 2,718 MW. Of these savings, 45% were due to utility programs, 37% to building code improvements, 16% to appliance efficiency standards, and the remainder to miscellaneous efforts (California Energy Commission 1986). Utilities can and should encourage these savings by supporting building code and efficiency standards, where such standards are cost-effective for most of their customers. For example, Puget Power was an active supporter of recent efforts to strengthen the Washington State building code. Issues likely to be debated in the 1990s include efficiency standards for lamps, lighting fixtures, and motors (Nadel 1990b); improved energy standards for residential and commercial buildings; and strengthened efficiency standards for fluorescent ballasts and residential appliances. Furthermore, by encouraging equipment purchases and construction practices that exceed code requirements, utility programs can make additional improvements to codes and standards possible. An example of this approach is BPA's residential new construction program.

Role of Utility-Operated Programs versus Bidding

Bidding programs offer an alternative delivery mechanism for C&LM services. While bidding can play an important role, the relationship between bidding programs and other utility C&LM efforts needs clarification. How does a utility decide whether to operate its own program or whether to sign a contract with a bidder? Should bidders and the utility both pursue conservation investments with the same customers? Addressing these issues can remove some of the uncertainties facing utilities and bidders, thereby allowing both utility and bidding programs to flourish.

How Much Savings Can Ultimately Be Achieved?

Estimates of the technical conservation potential are useful for establishing broad goals and research priorities, but the key question is how much energy can actually be saved through cost-effective programs and policies. Good estimates of achievable conservation potential for the utility sector are not presently available, due in considerable part to the shortage of good evaluation data. As additional evaluation data become available, including data from new and improved program offerings, estimates of achievable conservation potential should be made. These estimates can be of great use for utilities and other interested parties trying to identify program and planning priorities.

Acknowledgements

The research on programs for C&I customers summarized in this chapter was compiled as part of a project conducted for the New York State Energy Research and Development Authority, the New York State Energy Office, and Niagara Mohawk Power Corporation.

References

Anderson, K. 1990. "Appliance Efficiency Programs: Beyond Rebates." *Home Energy,* March/April, 13–17.

Anderson, K., and N. Benner. 1988. "The Energy Edge Project: Energy Efficiency in New Commercial Buildings." Paper presented to the American Society of Mechanical Engineers. Portland, Ore.: Pacific Power & Light.

Berry, L. 1989. *The Administrative Costs of Energy Conservation Programs.* ORNL/CON-294. Oak Ridge, Tenn.: Oak Ridge National Laboratory.

——. 1990. *The Market Penetration of Energy-Efficiency Programs.* ORNL/CON-299. Oak Ridge, Tenn.: Oak Ridge National Laboratory.

Blevins, R., and B. Miller. 1989a. *1988 Survey of Residential-Sector Demand-Side Management Programs.* EPRI CU-6546. Palo Alto, Calif.: Electric Power Research Institute.

——. 1989b. *1987 Survey of Commercial-Sector Demand-Side Management Programs.* EPRI CU-6294. Palo Alto, Calif.: Electric Power Research Institute.

BNL. See Brookhaven National Laboratory.

Brookhaven National Laboratory. 1987. *Analysis and Technology Transfer Annual Report 1986.* Upton, N.Y.: Brookhaven National Laboratory.

Brown, M., and G. Reeves. 1985. *The Implementation Phase of a Residential Energy Conservation Shared Savings Program: The General Public Utilities Experience.* ORNL/CON-187. Oak Ridge, Tenn.: Oak Ridge National Laboratory.

Brown, M., and D. White. 1987. *Impact Analysis of a Residential Energy Conservation Shared Savings Program: The General Public Utilities Experience.* ORNL/CON-217. Oak Ridge, Tenn.: Oak Ridge National Laboratory.

California Energy Commission. 1986. *Conservation Report.* Sacramento: California Energy Commission.

California Public Utilities Commission. 1987. *Standard Practice Manual for Cost-Benefit Analysis of Demand-Side Management Programs.* Sacramento: California Public Utilities Commission.

Cambridge Systematics. 1988. *Evaluation of the Commercial Audit Program. Final Report.* Portland, Ore.: Bonneville Power Administration.

CEC. See California Energy Commission.

Chernick, P., I. Goodman, and E. Espenhorst. 1989. *Analysis of Fuel Substitution as an Electric Conservation Option.* Report prepared for Boston Gas Company. Boston: PLC, Inc.

CLF. Conservation Law Foundation of New England. See New England Electric System and Conservation Law Foundation of New England.

Clinton, J., and A. Goett. 1989. "High Efficiency Fluorescent Lighting Program: An Experiment with Marketing Techniques to Reach Commercial and Small Industrial Customers." In *Proceedings for the 1989 Conference on Energy Conservation Program Evaluation: Conservation and Resource Management.* Argonne, Ill.: Argonne National Laboratory.

Collins, N., L. Berry, R. Braid, D. Jones, C. Kerley, M. Schweitzer, and J. Sorensen. 1985. *Past Efforts and Future Directions for Evaluating State Energy Conservation Programs.* ORNL-6113. Oak Ridge, Tenn.: Oak Ridge National Laboratory.

CPUC. See California Public Utilities Commission.

Dagang, D. 1990. "Impact Evaluation of the Commercial Incentives Pilot Program." In *Proceedings of the 1990 ACEEE Summer Study on Energy Efficiency in Buildings.* Washington, D.C.: American Council for an Energy-Efficient Economy.

DOE. See U.S. Department of Energy.

Ecker, L., and L. Michelsen. 1989. "Comparative Analysis of the Effect of Incentives on Weatherization Programs." Augusta, Maine: Central Maine Power.

Egel, K. 1986. *Final Report of the Santa Monica Energy Fitness Program.* City of Santa Monica, Calif.

EIA. See Energy Information Administration.

Energy Information Administration. 1989. *Annual Energy Review 1988.* DOE/EIA-0384 (88). Washington, D.C.: U.S. Department of Energy.

Estey, D. 1989. "Bidding Conservation Against Cogeneration: The Level Playing Field." In *Demand-Side Management Strategies for the 90s. Proceedings: Fourth National Conference on Utility DSM Programs.* EPRI CU-6367. Palo Alto, Calif.: Electric Power Research Institute.

Faruqui, A., M. Mauldin, S. Schick, K. Seiden, G. Wikler, and C. Gellings. 1990. *Efficient Electricity Use: Estimates of Maximum Energy Savings.* EPRI CU-6746. Palo Alto, Calif.: Electric Power Research Institute.

Faruqui, A., K. Seiden, and S. Braithwait. 1990. *Impact of Demand-Side Management on Future Customer Electricity Demand: An Update.* Palo Alto, Calif.: Electric Power Research Institute.

Flaim, T. 1990. Letter to author, January. Niagara Mohawk, Syracuse, N.Y

Gage, J., and J. Niewald. 1989. "This New House: An Evaluation of a Utility Sponsored New Home Program." In *Demand-Side Management Strategies for the 90s. Proceedings: Fourth National Conference on Utility DSM Program.* EPRI CU-6367. Palo Alto, Calif.: Electric Power Research Institute.

Geller, H. 1987. "Energy and Economic Savings from National Efficiency Standards." Washington, D.C.: American Council for an Energy-Efficient Economy.

Geller, H., and S. Nadel. 1989. "Electricity Conservation: Potential Vs. Achievement." Paper presented 11–13 September at the Least Cost

Utility Planning Conference of the National Association of Regulatory Utility Commissioners in Charleston, S.C.

Goldman, C., and E. Hirst. 1989. *Key Issues in Developing Demand-Side Bidding Programs*. LBL-27748. Berkeley, Calif.: Lawrence Berkeley Laboratory.

Goldstein, D. 1990. "Super Appliance Rebates: Incentives for the Development of New Appliance Efficiency Technologies." San Francisco: Natural Resources Defense Council.

Gustafson, G., and J. Peters. 1987. *Process Evaluation of the Industrial Test Program. Final Report*. Portland, Ore.: Bonneville Power Administration.

Harris, J., and R. Diamond. 1989. *Energy Edge Impact Evaluation. Findings and Recommendations from the Phase One Evaluation*. Berkeley, Calif.: Lawrence Berkeley Laboratory.

Hicks, E. 1989. "Third Party Contracting Versus Customer Programs for Commercial/Industrial Customers." In *Energy Conservation Program Evaluation: Conservation and Resource Management. Proceedings of the August 23–25, 1989, Conference*. Argonne, Ill.: Argonne National Laboratory.

Hirst, E. 1984. "Evaluation of Utility Home Energy Audit (RCS) Programs." In *Doing Better: Setting an Agenda for the Second Decade*. Vol. G, *Federal and State Programs*. Washington, D.C.: American Council for an Energy-Efficient Economy.

——. 1987. *Cooperation and Community Conservation*. Final Report. Hood River Conservation Project. DOE/BP-11287-18. Portland, Ore.: Bonneville Power Administration.

——. 1990. "Balancing the Scales with Demand-Side Data." *The Electricity Journal* 3 (May): 28–33.

Hirst, E., and P. Hu. 1983. *The Residential Conservation Service in Connecticut: Evaluation of the CONN SAVE Program*. ORNL/CON-132. Oak Ridge, Tenn.: Oak Ridge National Laboratory.

Krause, F., E. Vine, and S. Gandhi. 1989. *Program Experience and Its Regulatory Implications. A Case Study of Utility Lighting Efficiency Programs*. LBL-28268. Berkeley, Calif.: Lawrence Berkeley Laboratory.

Kreitler, V., and T. Davis. 1987. *High Efficiency Refrigerator Pilot Program—Final Analysis Report*. Binghamton, N.Y.: New York State Electric and Gas.

Kuenzi, C., and B. Wood. 1987. "Evaluation of Energy and Cost Savings from Replacement and Maintenance of Residential Air Conditioning Equipment. Salt River Project." Phoenix.

Kushler, M., P. Witte, and S. Ehlke. 1989. "Are High-Performance Residential Programs Still Feasible? The Santa Monica RCS Model Revisited." In *Energy Conservation Program Evaluation: Conservation and Resource Management. Proceedings of the August 23–25, 1989, Conference*. Argonne, Ill.: Argonne National Laboratory.

McRae, M., S. George, and M. Koved. "What are the Net Impacts of Residential Rebate Programs?" In *Proceedings of the 1988 ACEEE*

Summer Study on Energy-Efficiency in Buildings. Vol. 9, *Program Evaluation.* Washington, D.C.: American Council for an Energy-Efficient Economy.

Major, M., and B. Cody. 1987. "A Market Test of Solar and Heat Pump Water Heaters: Promotion vs. Incentives." In *Energy Conservation Program Evaluation: Practical Methods, Useful Results. Proceedings of the 1987 Conference.* Vol. 2. Argonne, Ill.: Argonne National Laboratory.

Miller, P., J. Eto, and H. Geller. 1989. *The Potential for Electricity Conservation in New York State.* NYSERDA 89–12. Albany: New York State Energy Research and Development Authority.

Morgan, S. 1990. "Moving Beyond the Weatherization Assistance Program: Energy Services for Low Income Customers." In *Proceedings of the 1990 ACEEE Summer Study on Energy Efficiency in Buildings.* Vol. 8, *Utility Programs.* Washington, D.C.: American Council for an Energy-Efficient Economy.

Morgan, S., and E. Katz. 1984. "The Financing of Energy Conservation Services to Low Income Households: Alternatives to Grants." in *Doing Better: Setting an Agenda for the Second Decade.* Vol. H, *Low Income and Equity.* Washington, D.C.: American Council for an Energy-Efficient Economy.

Moskovitz, D. 1989. *Profits and Progress through Least-Cost Planning.* Washington, D.C.: National Association of Regulatory Utility Commissioners.

Nadel, S. 1988. "Utility Commercial/Industrial Lighting Incentive Programs: A Comparative Evaluation of Three Different Approaches Used by the New England Electric System." In *Proceedings of the 1988 ACEEE Summer Study on Energy-Efficiency in Buildings.* Vol. 6, *Utility and Private Sector Conservation Programs.* Washington, D.C.: American Council for an Energy-Efficient Economy.

——. 1990a. *Lessons Learned: A Review of Utility Experience with Conservation and Load Management Programs for Commercial and Industrial Customers.* NYSERDA 90–8. Albany: New York State Energy Research and Development Authority.

——. 1990b. "Efficiency Standards for Lamps, Motors, and Lighting Fixtures." In *Proceedings of the 1990 ACEEE Summer Study on Energy-Efficiency in Buildings.* Vol. 7, *Government, Nonprofit, and Private Progams.* Washington, D.C.: American Council for an Energy-Efficient Economy.

Nadel, S., and M. Ticknor. 1989. "Electricity Savings from a Small C&I Lighting Retrofit Program: Approaches and Results." in *Proceedings for the 1989 Conference on Energy Conservation Program Evaluation: Conservation and Resource Management.* Argonne, Ill.: Argonne National Laboratory.

NAERC. See North American Electric Reliability Council.

NEES. See New England Electric System.

Neos Corp. 1989. "Final Report. Operating a Commercial Lamp Installation Program." Sacramento, Calif.: Western Area Power Administration.

New England Electric System. 1988a. *Evaluation Report on Massachusetts Electric Company's Enterprise Plan. Executive Summary.* Westborough, Mass.: New England Electric System.

——. 1988b. "Six-Month Evaluation of New England Electric System Partners in Energy Planning Programs." Westborough, Mass.: New England Electric System.

——. 1990. *Conservation and Load Management Annual Report.* Westborough, Mass.: New England Electric System.

New England Electric System and Conservation Law Foundation of New England. 1989. *Power by Design: A New Approach to Investing in Energy Efficiency.* Westborough, Mass.: New England Electric System.

North American Electric Reliability Council. 1989. *1989 Electricity Supply and Demand.* Princeton, N.J.: North American Electric Reliability Council.

Okumo, D. 1990. "Multifamily Retrofit Electric Savings: The Seattle City Light Experience." In *Proceedings of the ACEEE 1990 Summer Study on Energy Efficiency in Buildings.* Vol. 6, *Program Evaluation.* Washington, D.C.: American Council for an Energy-Efficient Economy.

Orange and Rockland. 1988. *End-Use Conservation Plan Results: 1987.* Pearl River, N.Y.: Orange and Rockland Utilities.

Pacific Gas and Electric Company. 1990. "Application of Pacific Gas & Electric Company for Authority to Adjust Its Gas Rates Effective January 1, 1991, to Implement an Expanded Customer Energy Efficiency Program Resulting from the Statewide Collaborative Process." San Francisco: Pacific Gas and Electric.

Peach, H. 1989. *Evaluation of Western Massachusetts Electric Company Residential Sector Performance Contracting Pilot Program.* Hartford, Conn.: Northeast Utilities.

Peters, J. 1989. *Interim Process Evaluation of the Bonneville Power Administration's Energy Savings Plan (E$P) Program.* Portland, Ore.: Bonneville Power Administration.

PG&E. See Pacific Gas and Electric Company.

Research Triangle Institute. 1989. *Residential Financial Incentives: Pilot Studies. Final Report.* Vol. 1, *Executive Summary.* Syracuse, N.Y.: Niagara Mohawk Power Corp.

Rogers, E. 1989. "Evaluation of a Residential Appliance Rebate Program Using Billing Record Analysis." In *Energy Conservation Program Evaluation: Conservation and Resource Management. Proceedings of the August 23–25, 1989, Conference.* Argonne, Ill.: Argonne National Laboratory.

Schick, S., I. Birnbaum, J. Blagden, and M. Adelaar. 1990. "Review and Assessment of U.S. Utility Experience with Residential Energy-Efficient Programs." Draft report submitted to Ontario Hydro by Barakat and Chamberlin, Inc., Oakland, Calif.

Schweitzer, M., M. Brown, and D. White. 1989. *Electricity Savings One and Two Years After Weatherization: A Study of 1986 Participants in Bonneville's Residential Weatherization Program.* ORNL/CON-289. Oakridge, Tenn.: Oakridge National Laboratory.

Shaffer, J., W. Adefris, B. Coates, D. Okumo, and D. Sumi. 1989. *Energy Conservation Accomplishments: 1977–1988.* Seattle: Seattle City Light.

Southern California Edison. 1983. *Conservation and Load Management: 1982 Program Results.* Rosemead, Calif.: Southern California Edison.

Stern, P., L. Berry, and E. Hirst. 1985. "Residential Conservation Incentives." *Energy Policy,* April, 133–42.

Stout, T., and W. Gilmore. 1989. "Motor Incentive Programs: Promoting Premium Efficiency Motors." Paper presented at the Electric Council of New England Demand-Side Management Conference in November, Boston.

Tempchin, R., A. Van den Berg, V. Geba, C. Felix, and M. Goldsmith. 1990. "Emission Impacts of Demand-Side Management Programs." In *Proceedings of the 1990 ACEEE Summer Study on Energy Efficiency in Buildings.* Washington, D.C.: American Council for an Energy-Efficient Economy.

Temple, Barker and Sloane, Inc. 1987. *Final Evaluation of the Western Massachusetts Electric Company's Performance Contracting Pilot Program.* Hartford, Conn.: Northeast Utilities.

Train, K., and J. Strebel. 1986. *Net Savings from the 1983 Audit and Hardware Rebate Programs for Commercial and Industrial Customers.* Vol. 1, *Summary.* Rosemead, Calif.: Southern California Edison.

U.S. Department of Energy. 1987. *1987 General and Summary Reports to Congress on the Residential Conservation Service Program.* Washington, D.C.: U.S. Department of Energy.

——. 1989. *Technical Support Document. Energy Conservation Standards for Consumer Products: Refrigerators and Furnaces.* DOE/CE-0277. Washington, D.C.: U.S. Department of Energy.

Vine, E., B. Barnes, and R. Ritschard. 1987. *Implementation of Home Energy Rating Systems.* LBL-22872. Berkeley, Calif.: Lawrence Berkeley Laboratory.

Vine, E., and J. Harris. 1988. *Planning for an Energy-Efficient Future: The Experience with Implementing Energy Conservation Programs for New Residential and Commercial Buildings.* LBL-25525. Berkeley, Calif.: Lawrence Berkeley Laboratory.

Vories, R. 1986. "Marketing the Austin Energy Star Program Using Private Sector Techniques to Market a Public Sector Program." In *Proceedings from the ACEEE 1986 Summer Study on Energy Efficiency in Buildings.* Vol. 5, *Marketing.* Washington, D.C.: American Council for an Energy-Efficient Economy.

Weedall, M., and F. Gordon. 1990. "Utility Demand-Side Management Program Incentive Programs: What's Been Tried and What Works to Reach the Commercial Sector." In *Proceedings of the 1990 ACEEE Summer Study on Energy Efficiency in Buildings.* Vol 8, *Utility Programs.* Washington, D.C.: American Council for an Energy-Efficient Economy.

Wirtshafter, R., and M. Koved. 1984. "Utility Conservation Programs Designed to Improve Participation Rates Among Low-Income Customers." In *Doing Better: Setting an Agenda for the Second Decade.* Vol. H. Washington, D.C.: American Council for an Energy-Efficient Economy.

Wisconsin Electric. 1989. *1988 Smart Money Energy Program Evaluation. Final Report.* Milwaukee: Wisconsin Electric Co.

Xenergy, Inc. 1990. *Final Report, Commercial Audit Pilot.* Binghamton: New York State Electric and Gas Corp.

Chapter 4

End-Use Load Shape Data Application, Estimation, and Collection

Joseph H. Eto and **Hashem Akbari,** *Lawrence Berkeley Laboratory*
Robert G. Pratt, *Pacific Northwest Laboratory*
Steven D. Braithwait, *Electric Power Research Institute*

As electric utilities increasingly adopt least-cost integrated resource planning processes, their information needs about demand-side management (DSM) options expand considerably. Many DSM alternatives offer the potential to meet a significant share of consumers' demands for energy services by means of increased energy efficiency (Krause and Eto 1988). However, comprehensive information on the cost and performance of these alternatives has been slow to develop. Current utility supply-side planning methods involve detailed assessments of alternative resource plans that take explicit account of the time-varying nature of customers' demands for electricity. In order for demand-side options to be treated comparably to generation resources, planners need reliable information on the impact of these options on system loads.

Yet information on the end-use components of aggregate electricity loads and how these components can be modified is not widely available, especially information for the time intervals used to evaluate generation options. A recent assessment of least-cost planning (LCP) concludes that uncertainty about the performance of DSM activities, including their impact on load shapes, is a major barrier to increased utility reliance on DSM for meeting customers' demands for electricity services (Goldman, Hirst, and Krause 1989).

The goal of this chapter is to assess progress in reducing this uncertainty. The chapter reviews leading efforts to collect, estimate, and apply end-use load shape data for utility planning purposes.

This chapter addresses utilities' need for load shape data from residential and commercial buildings (although other groups, such as the building energy research community, also need these data). We do not review industrial-sector end-use load shape data because efforts to develop these data—which are as important as residential and commercial end-use load shape data—are still in their infancy.

The first section of the chapter reviews utility applications of end-use load shape data; the next two sections review current efforts to obtain end-use load shape data by estimation and by direct metering. Our discussion of these three topics begins with a brief historical summary and a review of the state of the art. We also speculate on promising future areas of research. These speculations form the basis for a final section, which describes our vision of the next generation of end-use load shape data applications, estimation, and collection.

Utility Applications of End-Use Load Shape Data

End-use load shape data play a crucial role in several aspects of utility planning, including demand-side technology or program performance, load forecasting, and supply- and demand-side resource integration. The increased importance of these planning functions is the principal reason for current utility interest in acquiring end-use load shape data.

At the same time, these applications require different types of end-use load shape data. To forecast system load shapes for capacity planning, average load shapes by customer or rate class may be sufficient; needed information may include seasonal or weekly fluctuations and the hourly pattern of load shapes over a 24-hour period. For assessing a list of potential DSM options or developing a comprehensive integrated resource plan, planners may need additional load shapes for major end uses and even for specific technologies within those end uses. The precision required of this information will also depend on whether the analysis is a preliminary one or the final review prior to resource acquisition. Finally, a comprehensive analysis of the measured performance of a demand-side technology or program will require substantial data in addition to those on load shape in order to assign causal linkages between demand-side intervention and its measured consequences.

In this section, we review these three types of applications in order to better understand the motivation for efforts to estimate and collect end-use load shape data, efforts described in the following sections.

Demand-Side Management Applications of End-Use Load Shape Data

Some of the earliest applications of end-use load shape data can be found in early evaluations of DSM programs designed to modify utility load shapes. Examples include old reports by the Association of Edison Illuminating Companies' (AEIC) Load Research Committee examining the load shape impacts of marketing specific end uses, such as electric water-heating and air-conditioning (AEIC 1974). Applications also appear in research sponsored by the U.S. Department of Energy (DOE) and Electric Power Research Institute (EPRI) on the impacts of time-of-day electricity tariffs on various customer classes (Caves, Christensen, and Herriges 1984) and in DOE and EPRI evaluations of the aggregate impacts of utility direct load control programs.

More recently, it has become evident that planners must be aware of the impacts on load shape of all demand-side technologies and programs, not just those that address utility peak demands. For example, the Hood River Conservation Project, a landmark demonstration of the blitz approach for deploying residential DSM, aimed primarily at saving energy but also had measurable load shape impacts. The project involved the wholesale retrofitting of an entire community in Oregon. The retrofits stressed improvements to the thermal integrity of the homes. As part of the project, the electrical demands of 320 homes were separately monitored. The monitoring permitted researchers to quantify peak demand savings of more than 0.5 kW/household overall and more than 0.8 kW/household for electrically heated homes (Stovall 1989). More importantly, these peak demand savings, when combined with overall energy savings, indicated that household load factors had declined (a load factor is defined as the average demand divided by the peak demand). This observation led to the suggestions that further improvements could be made by downsizing home heating equipment in response to the reduced thermal loads.

Similarly, evaluation, as well as monitoring of demand-side technologies or programs, has been substantially refined by the availability of end-use data. For example, in evaluating the net impact of utility direct load control (DLC) programs, it is well known that the normal cycling behavior of controlled appliances may or may not be affected by a utility's program: that is, absent the program, significant cycling may already be the pattern of normal operation. Thus, the DLC program's impact on aggregate load shape is a function of how much the program has modified the distribution of appliance cycling times for a population of users. Evaluating these distributions requires large samples and data collection on days when the program is operating and on days when it is not. See Braithwait 1989 for a recent evaluation of a DLC program that takes explicit account of these distributions.

We expect that applications of end-use load shape data for DSM evaluation will assume increased importance as demand-side planning becomes more integrated into utility planning processes. In particular, the use of competitive resource acquisition mechanisms (such as demand-side bidding) to solicit demand-side resources from third parties will increase the need for explicit measurement of demand-side program savings. End-use load shape data should play a prominent role in these evaluations.

Forecasting Applications of End-Use Load Shape Data

Forecasting utility system load shapes through the summation of end-use load shapes is the logical consequence of utility adoption of the end-use framework for forecasting annual energy use. One of the earliest examples of this linkage is the system of forecasting models developed by the California Energy Commission (CEC). In response to a statutory charge to prepare forecasts of future energy use that explicitly capture the effects of California's building and appliance standards, CEC developed the first generation of end-use forecasting models for application to distinct utility service territories. The modeling system, which continues to be refined by the CEC, included an end-use peak demand model to forecast hourly system loads for separate end uses during a utility peak day (Jaske 1980). The model operates as a post-processor to forecasts of annual end-use electricity demands predicted by a separate model. The explicit linkage between the annual energy demand forecast and the system peak-day load shape ensures an important consistency between the forecasts for annual energy and peak demand that is often lacking when the two quantities are forecast separately.

The commercially available counterpart to the CEC's peak demand model is the Hourly Electric Load Model (HELM). HELM is a flexible load shape forecasting model that takes user-entered forecasts of annual energy sales at a user-selected level of disaggregation (for example, total system, customer class, end use) and monthly, daily, and hourly allocation factors, again at optional levels of detail (typical days, 8,760 hours), and produces a system load shape forecast (ICF 1985). Most utilities currently use HELM at the customer class level, although some utilities model weather-sensitive uses separately. Indeed, much utility interest in end-use load shape data is for monthly and annual aggregations of these data for energy forecasting or other purposes unrelated to load shape.

For the future, we expect forecasting applications for end-use load shape data to increase. For example, some analysts believe it may soon be possible to produce annual energy forecasts by aggregating hourly end-use load shape forecasts (Eto, Blumstein, and Jaske 1988). Forecasting energy use on an annual basis is largely a matter of convenience. Many important energy use decisions (such as the usage, as opposed to the purchase, of an energy-using durable good) take place at a much finer level of temporal disaggregation. These forecasting efforts will only proceed, however, after significant advances in our understanding of the causal factors influencing energy use over shorter time intervals.

Integrated Resource Planning Applications of End-Use Load Shape Data

A distinguishing feature of early applications of load shape data is that they were not fully integrated into the process of utility resource planning. Despite producing forecasts of hourly system loads for the peak day, for example, the CEC model passed a forecast of only total annual energy and peak demand to the resource integration planners. Similarly, for most evaluations of demand-side resources, the load shape impacts of specific demand-side resources are manually subtracted from aggregate system load shapes, a practice that may ignore interactive effects among end uses and other DSM programs.

An emerging application of end-use load shape data is better integration of demand-side resources into the utility planning process. From a purely mechanical standpoint, these improvements are exemplified by the recent availability of demand-side screening and integrated resource planning models, which facilitate the analysis of demand-side resources. From a more theoretical standpoint, end-use load shape data are beginning to play an extremely important role in extending demand-side planning into the realms of transmission and distribution (T&D) system planning and fuel-switching.

Demand-Side Screening Analysis

Detailed analysis of all available demand-side resources is inefficient because initially only a handful of demand-side options will be appropriate for serious consideration. Reducing the long list of available resources is called screening analysis. At this initial stage of the planning process, shortcuts are taken to facilitate rapid analysis of a large number of options. For example, rates of implementation and marginal costs of programs may be fixed regardless of the size of the demand-side intervention. At the same time, due to their importance for utility planning, the load shape characteristics of demand-side resources and time-differentiated marginal costs of electricity generation will often be retained for this stage of analysis. Models that support screening analyses using end-use load shape data include COMPASS (SRC 1989) and DSManager (EPS 1989).

Integrated Resource Planning Models

With a manageable list of demand-side resources identified for further analysis, the need to consider these resources on an equal footing with those on the supply side has led to a whole new class of planning models. These models, called integrated resource planning models, combine historically distinct modeling capabilities, such as load forecasting and production costing, into a single piece of software. While extensive calibration and coordination of data transfer with the more detailed stand-alone models for each modeling task are required, the ability of these new models to carry out an integrated analysis rapidly makes them extremely attractive for strategic planning. Well-known integrated planning models that feature end-use load shape data handling capabilities include UPLAN (Lotus Consulting Group 1986), MIDAS (Farber, Brusger, and Gerber 1988), and LMSTM (Decision Focus 1982). (See also Eto 1990 for an overview of issues associated with the use of demand-side screening and integrated resource planning models.)

Published examples of the application of end-use load shape data with an integrated resource planning model are rare, although many such studies exist as proprietary consultant reports or as parts of utility regulatory filings. A recent exception is Comnes et al. 1988, which used the LMSTM model to evaluate the cost-effectiveness of utility incentives to stimulate adoption of cooling thermal energy storage technologies for buildings.

T&D System Planning

Most integrated resource planning efforts by utilities consider demand-side options only as means to modify decisions on whether and how to expand generating capacity. The availability of end-use load shape data and geographically differentiated utility substation metering has led to the possibility of also deploying demand-side programs to avoid T&D investments. A recent study examined utility T&D planning and concluded that significant savings could be realized by targeting DSM to specific geographic locales where the avoided cost of T&D was high due to the imminent need to upgrade the capacity of substation distribution transformers (Rosenblum and Eto 1986).

Fuel-Switching

Similarly, an integrated resource planning process should also (but typically does not) consider fuel-switching as a means for meeting customer's demands for energy services. End-use load shape profiles of both electric and gas energy-using equipment can play important roles in these evaluations. At this time, we are aware of only one study that has begun to compare these profiles, focusing on residential appliances (Quantum Consulting 1989).

The Need for End-Use Load Shape Data

It is probably safe to say that the sophistication of utility applications for end-use load shape data (in particular, currently available software models) exceeds the quality and quantity of currently available data. We see no sign that this trend will end soon. Acquiring end-use load shape data, however, is an expensive undertaking with large differences in cost between end-use load metering and load shape data estimation. Therefore, the relevant economic question to which we now turn is, given the value of end-use load shape data for utility planning, what is the most cost-effective means for obtaining them?

Estimating End-Use Load Shapes

Prior to the recent wave of end-use metering projects, the only means for obtaining load shape data unique to local conditions was estimation. Estimation methods have historically relied on extensive and largely unverifiable engineering judgment. Indeed, concern over the reliability of these judgments has been a major impetus for the end-use metering efforts described later. However, increased collection of supplementary data, such as customer mail surveys and load research data, has led to a whole new generation of estimation methods. Moreover, the availability of end-use metered data provides, for the first time, the potential for validating the estimation methods. When validated, these methods offer the promise of producing end-use load shape data at a fraction of the cost of metering.

Historical Development of Estimation Methods

Traditional approaches to load shape estimation typically have used engineering simulations. The basic approach was to use available supplementary data, such as results of mail surveys, on a subset of the building population (single family dwellings, large offices, and so on). Engineering judgment was applied to these data to create a prototypical building whose energy-use patterns were considered representative of the subset of buildings being studied. An hourly building energy simulation program then produced the end-use load shapes.

The basic issue for this method, as for all estimation methods, is calibration. For the earliest efforts, calibration was only possible at an extremely high level of end-use and temporal aggregation (usually monthly total energy bills). Even then, because lighting and equipment loads were used only as inputs to the thermal simulations, calibrations typically estimated only the relative magnitudes, not the shapes, of these loads. In the absence of more detailed data, independent judgment as to the accuracy of simulated hourly load shapes has been largely a matter of faith. Indeed, many early load shapes developed by the above method exhibit the characteristic square shape that arises from the simulation of prototypes. (See, for example, Akbari et al. 1990 for a review of some of these studies.)

State-of-the-Art Estimation Methods

The passage of the Public Utilities Regulatory Policy Act of 1978 (PURPA) provided an unanticipated benefit for end-use load shape estimation. PURPA directed utilities to carry out detailed cost-of-service studies, based on hourly measurement of customer loads, for the purpose of reforming rate design. As a result, hourly whole-building load shape data have become widely available. The benefit for end-use load shape estimation lies in the fact that these data provide a control for reconciling estimates of hourly or even shorter-interval load shapes. Unfortunately, until recently little or no information on customer characteristics was collected along with the load research data. A number of utilities, seeing the value of load research data for other than cost-of-service applications, have begun to collect characteristics data and in some cases to expand the samples to represent market segments in addition to rate classes.

Researchers have used at least six distinct methods of estimating end-use load shape in order to utilize these data. The methods are (1) one-dimension application of the Stephan-Deming Algorithm, (2) the variance allocation approach, (3) the End-Use Disaggregation Algorithm, (4) the conditional demand approach,

(5) the bi-level regression approach, and (6) the Statistically Adjusted Engineering (SAE) approach. For purposes of exposition, it is useful to separate the methods that are primarily deterministic from those that are primarily statistical (although, as we shall see, this distinction breaks down for several of the methods).

The deterministic methods, which include methods 1 through 3, rely on exact reconciliation to an hourly control total, which is provided by the whole-building load research data. Of the three methods of which we are aware, reconciliation starts with an engineering simulation of the sort relied upon by the earliest load shape estimation methods. The later methods, however, typically rely on much more detailed information to develop the simulation input (thereby minimizing the extensive reliance on engineering judgment that characterized many early efforts). More importantly, they start with the assumption that an engineering simulation will not equal the measured, whole-building load shape. Each method differs in the manner by which the difference between the observed total and the sum of the initial, simulated estimates is allocated to constituent end uses.

The most straightforward allocation method, called the one-dimensional application of the Stephan-Deming Algorithm, is simple proration of the difference between the observed total and the sum of the simulated end uses based on the relative magnitudes of end uses in the original simulated estimates (SRC 1988). If, for example, there are only two end uses and one is simulated to be twice the size of the other, two-thirds of the difference between the simulated total and the control total is allocated to the larger end use. This approach has been used to estimate commercial sector end-use load shapes for the Southern California Edison Company. More flexible versions of this simple allocation have been implemented in the RELOAD software discussed in the next section.

Another allocation rule, called the variance allocation approach, involves prorating the difference between the simulated and control totals based on the observed statistical variation in the simulated end-use loads (Schon and Rodgers 1990). The rationale for this approach is that highly variable loads are more likely than relatively stable loads to diverge from simulation-based estimates of their magnitudes. (Of course, the magnitude of the observed variation is also related to the magnitude of the initial load.) This approach has been applied to a study of commercial buildings in the Florida Power and Light Company service territory.

A final deterministic reconciliation method, called the End-Use Disaggregation Algorithm, treats weather-sensitive end uses (cooling and heating) separately from other end uses (Akbari et al. 1988). An exact estimate of the weather-sensitive end use is first derived from a regression of the control totals provided by the whole-building load research data on temperature for each hour of the day. (An intercept for the weather-sensitive end use estimated from an analysis of the simulated end-use data is also included to account for non-weather-sensitive cooling or heating.) The allocation of any remaining differences between the simulated and control total takes place only after the weather-sensitive end use has been subtracted from the control total. The allocation is based on the magnitude of the initial simulated loads (as is done in the Stephan-Deming method), subject to continuity constraints placed on adjacent hours to reduce fluctuations from hour to hour. The motivation for this approach is the assumption that the correlation of measured whole-building loads to observed weather is superior to simulations for estimating weather-sensitive end-use load shapes. The approach has been used to develop end-use energy utilization intensities (EUIs) and load shapes for commercial buildings in the Southern California Edison service territory (Akbari et al. 1989).

Statistical methods, which include methods 4 through 6, represent another major approach for utilizing whole-building load shape data in developing load shapes. As with the deterministic methods, the principal aim is to reconcile selected explanatory variables with some control total. For the deterministic methods, the explanatory variables are taken from an engineering simulation so as to provide a physical basis for the reconciliation (that is, we are adjusting estimates of end-use loads to match an observed or estimated control total), and the reconciliation to the control total is exact. The statistical methods typically rely on regression techniques that correlate explanatory variables with the hourly control total. These variables need not all be physical, and the reconciliation to the control total is an approximate one usually expressed as a goodness of fit.

The earliest application of the statistical method to end-use load shape estimation—called the conditional demand approach—was a direct extension of the conditional demand techniques used to estimate annual energy utilization intensities, which express end-use energy use per unit of floor area, or unit energy consumption (UEC), which expresses energy use per appliance. The conditional demand approach is essentially a correlation analysis of the energy use of many separate premises, such as homes or offices, against the portfolio of energy-using equipment in each of these premises. The analysis seeks to determine the difference in observed load due to the presence of a given energy-using device, all other things being held equal. This difference is taken to be the energy contribution of the device. The technique was first applied to annual and monthly billing data (Parti and Parti 1980). With the availability of whole-building load shape data, the extension of the technique to an hourly time-step was an obvious one. Published applications of this approach include Hill 1982; Parti and Sebald 1984; and Aigner, Sorooshian, and Kerwin 1983.

Purely correlational methods for end-use load shape estimation can be criticized for ignoring (or making little explicit use of) known determinants of energy use (such as the influence of weather on heating and cooling loads). Recently, two very different methods for incorporating this knowledge within a statistical framework have been developed. In effect, these methods are hybrids of the engineering approaches that underlie the deterministic methods and the previously described statistical correlations.

The first method, called the bi-level regression approach, involves two levels of time-series and cross-section regression analyses (1986). In the first level, the hourly load of individual households is regressed both against weather-related variables and against sine and cosine functions, which capture daily, weekly, and seasonal periodicity in loads that are independent of weather. In the second level, the coefficients estimated in the first level (separately for each individual household) are regressed as a group against customer characteristics.

The second method, the SAE approach, is very close in spirit to the deterministic reconciliation methods (CSI/CAI/ADM 1985). First, an engineering simulation is developed to provide an initial estimate of end-use loads. (A more recent implementation of this approach incorporates metered end-use load shape data from a limited sample of premises as the initial estimate for selected end uses. See Caves, Windle, and Kendall 1988.) Next, the initial estimates are regressed against the control totals, which are averages of hourly energy use for typical days. The estimated coefficients can then be thought of as adjustment factors that reconcile the initial estimates to the control total. In other words, correlational analysis is used to perform the allocation of differences statistically, whereas, in the first three methods, the allocation is performed deterministically.

Deterministic and statistical estimation methods both exhibit desirable qualities for end-use load shape development. Deterministic methods rely on engineering simulations that provide a direct physical link between loads and their causes. The specificity of engineering simulations also facilitates subsequent planning analyses of the likely effect of introducing demand-side technologies. The price of such specificity is the cost of obtaining the detailed information required to develop an engineering simulation. Statistical methods are valuable because, unlike engineering simulations, they do account for behavioral dimensions. Physically identical structures will use energy differently because energy-use decisions are made by individuals, not buildings. To the extent that the explanatory variables are independent, exhibit variation across the sample, and, most importantly, are statistically significant, statistical techniques can capture these behavioral influences implicitly. However, because the physical underpinnings of energy use are suppressed, the resulting models of energy use may not be equally amenable to what-if types of analysis. Of course, from the more limited standpoint of end-use load shape data development, the issue is the accuracy of these methods and their costs relative to alternative methods of obtaining these data.

Validating End-Use Load Shape Data Estimates with Measured Data

The availability of end-use metered data provides, for the first time, the opportunity to validate end-use load shape estimation methods. However, efforts to use these data for this important task remain in their infancy; we are aware of only two studies, both examining only residential end uses, that have used end-use load shape data to validate estimates.

The first study focused not on end-use load shapes, per se, but on the integrated sum of the hourly values to an annual energy use total by end use (Pratt et al. 1990). In this study, metered residential end-use data from several metering projects were compared to estimates of these end uses developed by conditional demand and engineering studies. The conditional demand estimates were found to be in good agreement (a statistical difference of 10% at the level of annual energy use totals) with the metering studies for refrigerators, freezers, dryers, ranges, and central air-conditioning. Poor agreement was found for dishwashers (the conditional demand estimate was too high), hot water (too low), and space heating (too high, although the comparison is suspected to have been influenced by the weather normalization method applied to the various study results). The engineering estimates were found to be in good agreement with the metering studies for water heating, refrigeration, and clothes washing (clothes washing was not examined by the conditional demand studies), but were in poor agreement for space heating (too high), central air-conditioning (too high), ranges (too high), and dishwashers (too high).

A second study evaluated the accuracy of estimated load shapes using the SAE and the bi-level regression approaches described above (CSI/CAI/SSI 1989). The validation was performed using residential end-use metered data gathered by the Pacific Gas and Electric Company and an engineering simulation for each load. For the SAE approach, substantial improvement over the engineering load estimates was observed for the weather-sensitive end uses. For the non-weather-sensitive end uses, the SAE approach appeared to introduce errors to the engineering load estimates. Finally, the SAE weather-sensitive end-use loads were more accurate for average days than for peak days. For the bi-level regression approach, the most accurate loads were estimated for central air-conditioning and clothes drying, while the least accurate loads were those estimated for refrigerators and water heaters.

These validation studies suggest that, at this time, statistical end-use load shape estimation methods may be well suited for capturing scheduled, non-weather-sensitive end uses. More importantly, they substantiate the potential reliability of the estimation methods for obtaining end-use load shape data at costs far less than end-use metering. The lack of validation studies for the deterministic methods precludes conclusions at this time. Additional validation efforts for all the methods over a wider range of locations, building types (especially in the commercial sector), and end uses will be required before the methods can be regarded as complete substitutes for end-use load metering. However, as we shall describe, it is not clear that future load shape data development efforts should be faced with such an either/or decision.

End-Use Load Shape Data Collection

The most intuitively appealing approach for developing end-use load shape information is to collect the data directly by metering the desired end uses. Given the high cost of end-use metering (which should, but often does not, include the necessary costs of analysis following the collection of data), it has been impractical to carry out data collection for more than a small sample of the population. Efforts to reduce these costs and to increase the explanatory power of data already collected are the focus of future work.

Early Collection Efforts

The earliest efforts to collect end-use load shape data for utility planning date back to publications in the 1960s by the Load Research Committee of the AEIC. These studies of individual, predominantly residential loads, such as electric water-heating and air-conditioning, were performed in support of utility electricity marketing efforts (AEIC 1974). In the late 1970s and early 1980s, these studies were joined by a host of individual building metering studies that were typically parts of larger research studies on the performance of conservation technologies. (A good summary of many commercial sector projects can be found in Heidell, Mazzucchi, and Reilly 1984.)

The distinguishing feature of these early studies is that they did not focus on the statistical generalizability of the results (which, by definition, could not be generalized to larger populations in the case of individual building studies). In large part due to the high cost of direct metering but also due to the fact that incorporation of the results into a utility planning process was never envisioned as part of the research design, these studies are of secondary importance for most utility planning purposes. Where some statistical sampling procedures did enter into the research design, as was the case for some of the AEIC studies, the age of these studies, some of which are close to 30 years old, makes continued use problematic.

State-of-the-Art Collection Efforts

More recently, electric utilities, realizing the value of end-use load shape data for planning purposes, have engaged in a number of end-use load shape metering studies. What distinguishes these studies is that the samples are often large, and, more importantly, use of the results for utility planning is an explicit and major justification for the projects.

We have identified 27 recent end-use metering projects in the United States (see Table 4-1). The list of commercial sector projects is felt to be reasonably complete and includes several projects just getting under way; however, the list of residential projects may reflect biases due to the authors' location in the western part of the United States. We are not aware of any industrial sector end-use metering efforts involving sample sizes approaching those of the projects in Table 4-1.

The first four columns of Table 4-1 describe the sponsor of the project, the geographic area under study, the project name, and the customer sectors. Note that several sponsors have more than one project (or one project that covers multiple segments of the residential, multifamily, and commercial building sectors). Multiple projects by a given sponsor are a testament to the increased importance these sponsors place on the use of metered data for improving planning assumptions and estimates.

Table 4-1. Recent major end-use metering projects.

Sponsor	Geographic Area	Project Name	Sector[a]	Sample Type[b]	# of Bldgs.	EU/ Bldg.[c]	Total EUs	Protocol Type[d]	Quality Control[e]	Time Resol.	Dur. (Yrs.)	Status
Bonneville Power Administration	Hood R., Ore.	Hood River	RES	Retro/Stat	314	3	942	All EU-Sub	Limit	15 min.	5	Completed
	Pacific NW	RSDP	RES	Exp/Ctrl	422	3	1,266	All EU-Sub	Limit	Weekly	2	Completed
	Pacific NW	ELCAP-Base	RES	Statistical	288	8	2,304	All EU	Sumcheck	Hourly	5	Ongoing
	Pacific NW	-Case	RES	Special	56	8	448	All EU	Sumcheck	Hourly	5	Ongoing
	Pacific NW	-RSDP	RES/MF	Exp/Ctrl	155	8	1,240	All EU	Sumcheck	Hourly	6	Ongoing
	Seattle, Wash.	-Base	COM	Statistical	103	12	1,236	All EU	Sumcheck	Hourly	4	Ongoing
	Pacific NW	-CREUS	COM	Retrofit	40	12	480	All EU	Sumcheck	Hourly	4	Ongoing
	Pacific NW	Energy Edge	COM	Experimental	28	7	196	All EU	Sumcheck	Hourly		Start-up
Seattle City Light	Seattle, Wash.	CHEUS	COM	Retro/Stat	7	3	21	All EU-Sub	Limit	Hourly	6	Completed
Tacoma City Light	Tacoma, Wash.	Multifamily	MF	Exp/Ctrl	100	3	300	All EU	Sumcheck	Hourly	–	Start-up
DOE & EPRI	National	MRI	RES	Statistical	150	6	900	Select EU	?	Monthly	1	Completed
Pacific Gas & Electric	N. Calif.	AMP	RES	Statistical	750	3	2,250	Select EU	Visual	30 min.	2+	?
	N. Calif.	MYCE	COM	Statistical	45	5	225	All EU	Sumcheck	30 min.	–	Start-up
So. Calif. Edison	So. Calif.		RES	Special	124	4	496	Select EU	Visual	5 min.	2+	?
	So. Calif.		RES	Geographical	100	4	400	Select EU	Lim/Vis	5 min.	–	Start Up
	So. Calif.	RESA	COM	Geographical	53	4	212	All EU-Sub	Lim/Vis	15 min.	2	Ongoing
Sierra Pacific	Nev., E. Calif.	EIP-Res	RES/MF	Statistical	105	4	420	Select EU	Visual	?	?	?
		-Com	COM	Statistical	105	4	420	Select EU	Visual	?	?	?
Wisconsin Electric	Milwaukee		COM	Statistical	50	4	200	Select EU	Visual	15 min.		Start-up
Northeast Utilities	Connecticut		RES	Exp/Ctrl	250	4	1,000	Select EU	Visual	15 min.		Start-up
	Connecticut		COM	Exp/Ctrl	75	5	375	Select EU	Visual	15 min.		Start-up
Several utilities	Massachusetts	JUMP	RES	Statistical	28	3	84	Select EU	Visual	?	2+	?
Ariz. Public Service	Arizona	LCEP	RES	Exp/Ctrl	100	3	300	Select EU	Visual	15 min.	2	?
Gulf States Utilities Co.	?		RES	Special	232	4	928	All EU-Sub	Lim/Vis	?	?	?
Consolidated Edison	New York		RES	Statistical	396	2+	792+	Select EU	Visual	?	?	Completed
State of Texas	Texas		COM	Retrofit								Start-up
Penn. Power & Light	Pennsylvania		COM	Statistical	49	3	147	All EU-Sub	Lim/Vis	?	1	Completed

[a] Sectors abbreviations: RES = Residential; MF = Multifamily; COM = Commercial. [b] Sample Type Abbreviations: Retro/Stat = Retrofit/Statistical; Exp/Ctrl = Experimental and Control (see text). [c] Average number of end uses per building (approximate): one may be by end use by subtraction for All EU-Sub. [d] End-use protocol type: All EU = separately metered; All EU-Sub = one by subtraction; Select EU = selected end uses/appliances only (see text).
[e] Quality control abbreviation: Lim/Vis = Building total limit and visual reasonableness checks (see text)

Column five indicates the type of sample design used. Recall that the applications described in the first section provided two primary motivations for the designs of metered samples: (1) characterization of the building population for planning and forecasting purposes and (2) evaluation of the impacts of specific demand-side technologies or programs. Statistically based sample designs are generally used to obtain data in support of planning and forecasting processes. These designs are based on customer billing or survey data so that the metered buildings can be analyzed as representing a larger population. The use of metering to support evaluations of individual technologies or programs is typically based on a nonrandom sample of participants in the program, although some retrofit programs have relied on statistical sampling procedures. For new building programs, the samples are almost always a somewhat arbitrary (statistically speaking) set of experimental buildings from a pilot test of the program. Some of these projects will also include a parallel set of newly constructed buildings representing current practice as controls for the experiment. Other sample types indicated in Table 4-1 include studies of buildings selected for specific reasons such as high consumption or presence of particular appliances (termed Special) and studies that seek to capture geographical diversity within a region (termed Geographical). In general, these latter two sample types do not formally incorporate statistical sampling procedures.

Columns six, seven, and eight indicate the scope of the projects as measured by the number of buildings, average number of end uses per building, and total number of end uses metered.

Columns nine and ten indicate the monitoring protocol and primary method used for quality control by each project. The protocols used to define end uses are split into three groups: (1) those in which all defined end uses and a separate building total are metered (All EU), (2) those that meter at least the total and the major end uses but obtain the remainder by subtraction (All EU-Sub), and (3) those that meter only selected appliances or end uses in each building (Select EU). Among other things, the protocols determine the quality control procedures that may be applied. These procedures include a continuous energy balance using the building total as a sum-check, limit checks against monthly utility bills (if the remainder end use is small relative to the total consumption), and visual reasonableness and continuity checks when only selected end uses are metered.

Column eleven indicates the level of aggregation of the metered data. The time resolution of the data is typically 5- or 15-minute intervals for regions where peak loads are the central planning issue, and hourly where annual energy is the primary concern (the Pacific Northwest).

Finally, columns twelve and thirteen indicate the duration of the projects and their current status if known.

Metering end uses for a large number of buildings is expensive. Costs depend on the level of detail called for by the measurement protocol and on whether economies of scale can be realized with a given sample size. Fully burdened costs for large, detailed, all end-use protocol projects including sum-check quality control procedures and a duration of two years are currently in the range of $15,000–$25,000 per commercial building and $3,000–$7,000 per residential building. The cost is split about equally between installation and maintenance, with the installation costs split about equally between hardware and labor. Importantly, these costs do not include the considerable effort required to develop software in order to archive and analyze the data.

These costs also mean that despite the explicit reliance on statistical techniques in some of the sample designs, the final samples are often not very large for a given stratum. As a consequence, the resolution of the analyzed data is often not very precise. For example, it is not uncommon to find that standard deviations across buildings for a given end use are equal to or greater in magnitude than the observed means. Statistically speaking, this means that the null hypothesis of the measured load being equal to zero cannot be rejected at the 95% confidence level!

Lowering the Costs of End-Use Load Shape Data Collection

The desire for increased statistical precision in end-use load shape estimates calls for research in three technical areas of end-use load shape data collection: sample size determination, project duration, and metering costs per end use. This need also justifies seeking less expensive means for obtaining end-use load shape data, such as the estimation techniques described in the previous section and, as we shall describe, the use of end-use load shape data collected by others (data transfer).

We are not aware of specific studies examining the issue of increased data precision as a function of sample size. We note from our experience, however, that mean end-use loads tend to stabilize with sample sizes of about 20. Nevertheless, even larger sample sizes may be required to explain observed variances. For example, some researchers have observed that some causal relationships, such as the effect of number of occupants on water heating loads, can be obscured by other sources of variance when the sample size falls below 20 (Pratt et al. 1990). At the same time, other variance reduction techniques are possible. Others have suggested that it is possible to link small end-use metered samples with larger, whole-building load research samples to increase sample sizes, thus reducing variance, and thereby make end-use load estimates from very small samples representative of larger populations (Wright and Richards 1989).

On the issue of reducing the duration and cost of metering projects, there is evidence that the seasonal variation in nearly all residential and many commercial end uses (+10% to 20% of the mean) will preclude metering periods of less than a year from producing accurate results for some end uses (Pratt et al. 1990; Taylor and Pratt 1989). On the other hand, for some highly scheduled, non-weather-dependent end uses, such as commercial lighting and water heating, shorter duration metering periods may be warranted. At the same time, it should be recalled that the fixed costs of installing metering equipment are roughly half the total costs of metering (excluding analysis of the data) and that multiyear data also allow for study of price elasticity, occupancy and behavioral changes, retrofits and equipment changes, and persistence of savings from demand-side measures, among other topics.

More recently, efforts have been made to reduce the direct costs of metering end uses. One approach involves the use of decomposition techniques that track total electricity consumption at an extremely high level of time-resolution (about a thousandth of a second) in order to capture the signature of individual pieces of equipment turning on and off (Jones and Flagg 1989). Separate end-use loads are automatically detected by this decomposition, which in effect reduces the number of metering points per building to one. Another extremely promising approach involves the use of existing energy management systems as a direct source of equipment operating profiles (Flora, LeConiac, and Akbari 1986). (We also refer the reader to Harry Misuriello's review in this book of the state of the art in building energy performance monitoring.)

Using End-Use Load Shape Data Collected by Others

Perhaps the least expensive means for obtaining metered, end-use load shape data lies not with optimized sampling designs and better hardware, but with the transfer of results from existing metering studies. Prior to the recent era of end-use metering, which began around 1984, almost all utility applications of end-use load shape data relied on either secondary or estimated data. Attitudes were pragmatic: some metered data, from any source, was considered better than none simply because one did not have the ability to judge these data independently.

In addition to the grey literature of utility reports on individual metering projects (see the references for a selective listing of the reports underlying the metering studies reported in Table 4-1), we are aware of few published end-use load shape data compilations. Notably, the Bonneville Power Authority (BPA) has produced two major compilations of end-use load shape data from its ELCAP project, one for the residential and one for the commercial buildings being metered (Pratt et al. 1989; Taylor and Pratt 1989). In addition, researchers have compiled and analyzed residential end-use load shape data collected by California utilities in order to provide inputs for the California Energy Commission's peak load forecasting model (Ruderman et al. 1989). Finally, there is a relatively new software package, RELOAD (SRC/LCA/BCD 1988), which is distributed with a library of end-use load shapes that have been drawn from a number of sources, including engineering simulation and end-use metering.

Today, the increasingly extensive geographic coverage of end-use metering projects suggests that an adequate range of climatic and cultural diversity may nearly exist to characterize the residential sector for most of the United States and that this will also soon be achieved in the commercial sector. However, the widespread availability of these data will hinge on resolution of two important institutional and analytical issues.

Institutionally, there remains the need to establish mechanisms for equitably sharing and promoting the use of this expensive resource. An important issue is the confidentiality and propriety of data from donor utilities. Currently an informal questionnaire is being circulated to potential users and suppliers of data to help define parameters for some form of institutional data sharing (BPA 1990).

Analytically, substantial issues regarding the transferability of data remain unaddressed. These issues include climate normalization, control for regional structural characteristics, and control for occupant characteristics. We presume that these factors are responsible for a large part of the observed variability of end-use load shapes, along with a number of as yet to be determined random factors. Analysis to determine the nature of the influence of these factors on load shapes, leading to methods to adjust and transfer load shapes, is straightforward conceptually but complex in practice.

Considerable demonstration will be required to convince potential users that transferability is possible.

Two recent studies have begun to explore some of the prospects for load data transferability by comparing end-use load shape data from more than one metering study. The first study examined residential hourly load shape data aggregated to annual totals (or UECs) from four sources (Pratt et al. 1990). These studies found reasonable agreement for refrigeration, freezing, clothes washing, and hot-water use; this result suggests a certain consistency in behavior in the use of these appliances for these end uses. Larger differences were found for space heating, central air-conditioning, clothes drying, ranges, and dishwashing.

A second, previously mentioned study compared metered end-use load shapes from recent residential end-use metering projects sponsored by California utilities (Ruderman et al. 1989). End uses were first normalized by total energy use so that the comparison focused solely on differences in load shapes (not annual UECs). Not all end uses were metered in each project, but for the non-space-conditioning end uses metered by more than one project (including refrigeration, cooking, clothes drying, and water heating), very good agreement was observed for normalized load shapes. Air-conditioning load shapes (room and central air-conditioning) were first transformed into a matrix relating energy use to time of day. Temperature-humidity indices and load shapes were produced for peak days in several California climate zones. The normalized load shapes again showed very close agreement with one another.

Toward the Next Generation of End-Use Load Shape Data Development

We have reviewed recent efforts to apply, estimate, and collect end-use load shape data. Substantial progress has been made from the days when utility planning relied on independent (non-end-use) forecasts of peak load and energy use, and generally treated these future energy demands as immutable. Today, the value of end-use load shape data to inform and thereby improve utility planning decisions is largely undisputed. (See Gellings and Swift 1988 for an example of how this value can be quantified.) Progress in the application, collection, and estimation of end-use load shape data will be rapid. In this section, we sketch our observations on some of these directions.

Historically, potential applications have always led efforts to develop end-use load shape data. End-use load forecasting was well on its way toward becoming accepted utility planning practice several years before the initiation of recent end-use metering projects. We suspect that as end-use planning methods mature the need for end-use load shape data will increase. In particular, we believe that increased utility reliance on demand-side management options will only come about through increased utility confidence in the performance of these options.

We believe that increasingly sophisticated end-use planning approaches will also call for new types of end-use data. For example, historic segmentation based on building structural and operational characteristics (such as single-family residences or large offices) was based on a physical characterization of the ways in which energy is used. Shifting to a more behavioral characterization will call for load shapes segmented along very different lines, with end-user categories such as income level, form of property ownership, or membership in a needs-based market segment. (See NA/SRC/QEI 1989 for a description of this approach.) In this case, the planning needs are not for a new definition of end use but for increased supplementary data collection efforts that will improve our understanding of the causal factors underlying existing end uses.

As noted above, in the absence of metered data, analysts and planners have been forced to develop load shape information through assumptions about usage patterns, engineering simulation, statistical disaggregation of whole-building loads, and various combinations. A certain level of uncertainty exists with each of these techniques because they could not be validated without some end-use metered data. For this reason, all end-use load shape data users owe a debt of gratitude to the pioneering efforts by a handful of utilities to undertake comprehensive, well-designed end-use metering projects. The data from these projects offer the potential for extremely valuable data validation and leveraging efforts. These could take the form of adjustment techniques for transferring the load data from one location to another, testing and validation of various load shape estimation methods, and incorporation of new information into those same estimation methods.

In the near future, end-use data development techniques that leverage, or combine, features of engineering knowledge, prior end-use metering results, and statistical analysis of survey, billing, and load research data should be able to provide reliable, cost-effective information on end-use load shapes. Moreover, they will do so at a cost far less than that of large end-use metering projects. This scenario would suggest that future end-use metering projects be designed in close coordination with a comprehensive end-use load shape development effort. It might consist of, first, metering targeted at market segments or technologies for which little or no data are available and, second, small end-use metering samples designed to be leveraged with less expensive survey and whole-building load data. For end-use data development efforts to reach this level of maturity, a number of activities must take place. These involve important synergisms, and would benefit greatly by proceeding jointly. The following three general recommendations illustrate the type of process that could take place over the next few years.

First, existing end-use metering projects should soon provide adequate coverage of the most important building types, end uses, and geographic regions. Efforts to more fully exploit these data sources should be a high priority for future research. The primary goal of these efforts should be to develop the analytical procedures necessary to permit meaningful transfer of load shape data from one utility service territory to another. The procedures must explicitly capture the causal relationships underlying observed load shapes in order to control for differences in climate, building characteristics, and occupant behavior between service territories. These analytical efforts should proceed in parallel with institutional efforts to facilitate data transfer in which issues of confidentiality and propriety of data from donor utilities must be addressed.

Second, end-use load shape estimation methods should be able to produce data of sufficient accuracy for utility planning purposes. In particular, there is great promise in the use of hybrid estimation methods, which combine the best aspects of simulations, statistical analyses, and measured data. Efforts to utilize recent end-use metered data to validate estimation methods should be given the highest priority for research. From the standpoint of improving the estimation methods, a major challenge lies in determining the optimal amount of non-load shape data collection needed to support load shape estimation.

Third, it is likely that the realization of these two research objectives, driven by increased utility applications for end-use load shape, will still call for end-use load shape metering efforts. These incremental efforts will be a healthy sign for load shape development efforts if coordinated with load shape data transfer and improvements in estimation methods.

The costs of developing load shape data for utility planning can be significant, ranging in descending order from end-use metering to estimation to data transfer. Yet the benefits from improved resource planning will easily outweigh these costs. The issue for future end-use load shape development is not one of whether, but of how. We believe society will be best served when these efforts incorporate all potential sources of data including metering studies from other service territories, estimation based on utility-specific data, and local end-use metering. The challenges for future research lie in determining a cost-effective mix of these sources, not in choosing one over the other.

Acknowledgements

The study described in this chapter was funded by the Assistant Secretary for Conservation and Renewable Energy, Office of Utility Technologies, U.S. Department of Energy, under Contract No. DE-AC03-76SF00098.

References

AEIC. See Association of Edison Illuminating Companies.

Aigner, D., C. Sorooshian, and P. Kerwin. 1983. "Conditional Demand Analysis for Estimating Residential End-Use Load Profiles." University of Southern California, Los Angeles. Photocopy of working paper.

Akbari, H., K. Heinemeier, P. LeConiac, and D. Flora. 1988. "An Algorithm to Disaggregate Commercial Whole-Building Hourly Electrical Loads into End Uses." In *Proceedings from the ACEEE 1988 Summer Study on Energy Efficiency in Buildings.* Vol. 10, *Performance Measurement and Analysis.* Washington, D.C.: American Council for an Energy-Efficient Economy.

Akbari, H., J. Eto, J. Turiel, I. Heinemeier, K. Lebot, B. Nordman, and L. Rainer. 1989. *Integrated Estimation of Commercial Sector End-Use Load Shapes and Energy Intensities.* LBL-27512. Berkeley, Calif.: Lawrence Berkeley Laboratory.

Akbari, H., I. Turiel, J. Eto, K. Heinemeier, B. Lebot, and L. Rainer. 1990. "A Review of Existing Energy Use Intensity and Load-Shape Studies." In *Proceedings from the ACEEE 1990 Summer Study on Energy Efficiency in Buildings.* Vol. 3, *Commercial Data Designs and Technologies.* Washington, D.C.: American Council for an Energy-Efficient Economy.

Association of Edison Illuminating Companies. 1974. "Bibliography of Load Research Committee Reports." New York. Photocopy.

Bonneville Power Administration. 1990. "End-Use Data Interest Survey." Portland, Ore.: Bonneville Power Administration.

BPA. See Bonneville Power Adminstration.

Braithwait, S. 1989. "Measuring Direct Load Control Impacts." *EPRI Journal,* December.

Brodsky, J., and S. McNicoll. 1987. *Residential Appliance Load Study, 1985–1986.* San Francisco: Pacific Gas and Electric Company.

Cambridge Systematics, Christensen Associates, and ADM Associates. 1985. *Combining Engineering and Statistical Approaches to Estimate End-Use Load Shapes.* Vols. 1 and 2. EPRI-EA-4310. Palo Alto, Calif.: Electric Power Research Institute.

Cambridge Systematics, Christensen Associates, and Scientific Systems. 1989. *End-Use Load Shape Estimation: Methods and Validation.* Palo Alto, Calif.: Electric Power Research Institute.

Caves, D., L. Christensen, and J. Herriges. 1984. *Residential Response to Time-of-Use Rates.* 4 vols. EA-3560, EPRI-EQ-3560. Palo Alto, Calif.: Electric Power Research Institute.

Caves, D., R. Windle, and D. Kendall. 1988. "Residential End-Use Load Shapes: A Case Study." Paper presented at Demand-Side Management Symposium: Managing the Shape of Tomorrow on May 3–5 in Albany, N.Y. Palo Alto, Calif.: Electric Power Research Institute.

Comnes, G., E. Kahn, C. Pignone, and M. Warren. 1988. "An Integrated Economic Analysis of of Commercial Thermal Energy Storage." *IEEE Transactions on Power Systems* 3 (4): 1717–22.

CSI/CAI/ADM. See Cambridge Systematics, Christensen Associates, and ADM Associates.

CSI/CAI/SSI. See Cambridge Systematics, Christensen Associates, and Scientific Systems.

Decision Focus. 1982. *Load Management Strategies Testing Model.* EPRI-EA-2396. Palo Alto, Calif.: Electric Power Research Institute.

Electric Power Software. 1989. *DSManager User's Guide.* EPRI-CU-6564-CCML. Palo Alto, Calif.: Electric Power Research Institute.

EPS. See Electric Power Software.

Eto, J. 1990. "An Overview of Integrated Resource Planning Models." *Energy, the International Journal* 15 (11): 969–977.

Eto, J., C. Blumstein, and M. Jaske. 1988. "Changing Needs in Electricity Demand Forecasting: A California Perspective." In *Proceedings from the ACEEE 1988 Summer Study on Energy Efficiency in Buildings.* Vol. 8, *Planning and Forecasting.* Washington, D.C.: American Council for an Energy-Efficient Economy.

Farber, M., E. Brusger, and M. Gerber. 1988. *Multiobjective Integrated Decision Analysis System (MIDAS).* Vol. 1, *Model Overview.* EPRI-P-5402. Palo Alto, Calif.: Electric Power Research Institute.

Flora, D., P. LeConiac, and H. Akbari. 1986. "Methods to Obtain Building Energy Performance Data: Are Energy Management Systems Promising Sources?" In *Proceedings from the Energy Technology Conference.* Washington, D.C.: N.p.

Gellings, C., and M. Swift. 1988. "The Value of Load Research." *Public Utilities Fortnightly* June.

Goldman, C., E. Hirst, and F. Krause. 1989. *Least-Cost Planning in the Utility Sector: Progress and Challenges.* LBL-27130. Berkeley, Calif.: Lawrence Berkeley Laboratory.

Heidell, J., R. Mazzucchi, and R. Reilly. 1984. "Development of a Data Base on End-Use Energy Consumption in Commercial Buildings."

In *Proceedings from the ACEEE 1984 Summer Study on Energy Efficiency in Buildings*. Vol. D, *New and Existing Commercial Buildings*. Washington, D.C.: American Council for an Energy-Efficient Economy.

Hill, D. 1982. "The Time-of-Day Demand for Electricity by End Use: An Analysis of Wisconsin Data." In *Analysis of Residential Response to Time-of-Day Prices*. EPRI-EA-2380. Palo Alto, Calif.: Electric Power Research Institute.

ICF. 1985. *The Hourly Electric Load Model (HELM)*. Vol. 1, *Design, Development, Demonstration*. EPRI EA-3698. Palo Alto, Calif.: Electric Power Research Institute.

Jaske, M. 1980. "Analysis of Peak Load Demand Using an End-Use Load Forecasting Model." In *Proceedings of the EPRI End-Use Models and Conservation Analysis Workshop*. EPRI EA-2509. Palo Alto, Calif.: Electric Power Research Institute.

Jones, R., and D. Flagg. 1989. "A Preliminary Qualification Testing of EPRI's Non-Intrusive Load Monitor." In *Proceedings from the Information and Automation Technology Serving Electric Utility Customers in the 1990s Conference*. EPRI CU-6400. Palo Alto, Calif.: Electric Power Research Institute.

Krause, F., and J. Eto. 1988. *Least-Cost Utility Planning, a Handbook for Public Utility Commissioners*. Vol. 2, *The Demand-Side: Conceptual and Methodological Issues*. Washington, D.C.: National Association of Regulatory Utility Commissioners.

Lotus Consulting Group. 1986. *UPLAN Reference Manual*. Los Altos, Calif.: USAM Center.

NA/SRC/QEI. See National Analysts, Synergic Resources Corporation, and QEI.

National Analysts, Synergic Resources Corporation, and QEI. 1989. *Residential Customer Preference and Behavior: Market Segmentation Using CLASSIFY*. EPRI-EM-5908. Palo Alto, Calif.: Electric Power Research Institute.

Parti, M., and C. Parti. 1980. "The Total and Appliance-Specific Conditional Demand for Electricity in the Household." *The Bell Journal of Economics*, Spring, 309–21.

Parti, M., and A. Sebald. 1984. "Integrated Load and Time of Day Models for Electricity in Residences." In *Proceedings from the ACEEE 1984 Summer Study on Energy Efficiency in Buildings*. Washington, D.C.: American Council for an Energy-Efficient Economy.

Pratt, R., C. Conner, E. Richman, K. Ritland, W. Sandusky, and M. Taylor. 1989. *Description of Electric Energy Use in Single-Family Residences in the Pacific Northwest*. DOE/BP-13795-21. Portland, Ore.: Bonneville Power Administration.

Pratt, R., C. Conner, B. Cooke, and E. Richman. 1990. "Metered End-Use Consumption and Load Shapes from the ELCAP Residential Sample of Existing Homes in the Pacific Northwest." *Energy and Buildings*.

Quantum Consulting. 1989. *Residential Appliance End Use Survey:*

Collection of Residential Appliance Time of Use Energy Load Profiles, 1987/1988., Final Report. Rosemead, Calif.: Southern California Edison Company.

——. 1990. *Residential Energy Usage Comparison Project: An Overview*. EPRI-RP-2863-3. Palo Alto, Calif.: Electric Power Research Institute.

Rosenblum, B., and J. Eto. 1986. *Utility Benefits from Targeting Demand-Side Management Programs at Specific Distribution Areas*. EPRI-EM-4771. Palo Alto, Calif.: Electric Power Research Institute.

Ruderman, H., J. Eto, K. Heinemeier, A. Golan, and D. Wood. 1989. *Residential End-Use Load Shape Data Analysis*. LBL-27114. Berkeley, Calif.: Lawrence Berkeley Laboratory.

Schon, A., and R. Rodgers. 1990. "An Affordable Approach to End Use Load Shapes for Commercial Facilities." In *Proceedings from the ACEEE 1990 Summer Study on Energy Efficiency in Buildings*. Washington, D.C.: American Council for an Energy-Efficient Economy.

Scientific Systems. 1986. *Residential End-Use Load Shapes*. EPRI-EM-4525. Palo Alto, Calif.: Electric Power Research Institute.

SRC. See Synergic Resources Corporation.

SRC/LCA/BCD. See Synergic Resources Corporation, Laurits R. Christensen Associates, and Battelle-Columbus Division.

Stovall, T. 1989. "Load Shape Impacts of the Hood River Conservation Project." *Energy and Buildings* 13: 31–37.

Synergic Resources Corporation. 1989. *SRC/COMPASS Comprehensive Market Planning System, Users Guide*. Bala Cynwyd, Penn.: Synergic Resources Corporation.

Synergic Resources Corporation, Laurits R. Christensen Associates, and Battelle-Columbus Division. 1988. *DSM Customer Response*. Vol. 1, *Residential and Commercial Reference Load Shapes and DSM Impacts*. EPRI-EM-5767. Palo Alto, Calif.: Electric Power Research Institute.

Taylor, Z., and R. Pratt. 1989. *Description of Electric Energy Use in Commercial Buildings in the Pacific Northwest*. DOE/BP-13795-22. Portland, Ore.: Bonneville Power Administration.

Wright, R., and D. Richards. 1989. "Developing Information for Commercial Sectors Through Direct Metering." In *Proceedings of the End-Use Load Information and Application Conference*. Syracuse, N.Y.: The Fleming Group.

Chapter 5

Field Monitoring of Energy Performance in Buildings

Harry Misuriello, *The Fleming Group*

Field monitoring of energy performance is increasingly employed to obtain data required for making decisions about energy and power use in commercial and residential buildings. Data from field monitoring studies are used by energy suppliers, energy end users, building systems designers, public and private research organizations, equipment manufacturers, and public officials involved with the regulation of energy use in residential and commercial buildings. Although they may not address the same issues, these constituencies face the common challenge of obtaining accurate, reliable, and cost-effective performance data that meet specific information needs.

There are at least two reasons why the professional community interested in the energy efficiency of buildings needs to understand the nature and status of field monitoring. First, field monitoring, nearly always a complex and expensive undertaking, is usually funded either by government agencies using tax dollars or by utilities (and their research organizations) using revenue supplied by ratepayers; these sponsors have a responsibility to use public funds prudently. Second, project sponsors, data analysts, and data end users should understand the current capabilities and limitations of field monitoring methods in order to assure that their specific data needs are met.

This chapter reviews the state of the art of the field monitoring, defined as the gathering, with special-purpose recording equipment, of time-series field data on the quantity and patterns of energy use by buildings. Methods involving collection and analysis of normal utility billing records—an activity some refer to as "tracking" rather than "monitoring"—are not within the scope of this chapter.

Topics covered here are organized as follows:

1. *Types of monitoring projects.* In this review, monitoring projects are categorized into two families: utility load research monitoring and issues-research monitoring. These two families have different roles, missions, and constituencies and have little interaction. This chapter focuses on issues-research monitoring, a broad range of efforts to understand energy use in buildings for the purpose of promoting energy efficiency, energy conservation, and demand-side management (DSM).
2. *Forces that have affected the development of field monitoring.* The national energy legislation of the mid- to late 1970s and its resulting data requirements are analyzed as developmental influences. The evolution and current status of load research and issues research monitoring are compared and contrasted in the context of these forces.
3. *The state of the art of field monitoring for issues research.* This discussion constitutes the bulk of the chapter and focuses on three major elements that contribute to the effectiveness of field monitoring projects: monitoring protocols for project planning and specification, field data acquisition system (FDAS) hardware reliability and performance, and data quality assurance and control procedures. Most of the chapter is devoted to these three major topics because the quality of a project depends upon how well these elements are integrated.

The chapter then concludes with a discussion of future directions for field monitoring and a summary of the key points made.

Types of Field Monitoring Projects

Various research organizations have surveyed and catalogued a wide variety of projects involving field monitoring of energy performance in buildings (Heidell, Mazzucchi, and Reilly 1985; EPRI 1985). These studies have examined general types of monitoring projects with respect to level of detail, areas of research, requirements of data users, means of data collection, performance factors typically reported, and specific technologies under investigation. On the basis of these studies, two families of field monitoring activities encompassing four types of projects can be established. As shown in Table 5-1, monitoring projects can be broadly categorized in terms of goals and objectives, experimental approach, and level of monitoring detail. This chapter groups field monitoring projects into two families as follows:

Utility load research is generally concerned with aggregate energy use in buildings. Load research projects usually measure and characterize the average energy use of a representative sample of buildings or building systems. These projects typically have a small number of data points—usually one to four—in each building and use statistical analyses of a large number of buildings to provide results with the desired level of precision. Typical applications include cost-of-service studies, customer class-load studies, and rate design research. Load research data are also used to support regulatory reporting requirements and utility planning activities, as discussed in the next section.

Issues research monitoring is oriented toward understanding the performance of specific (but not necessarily typical) buildings, systems, and component technologies. Issues research monitoring typically addresses energy efficiency improvements, energy conservation, and DSM. Issues-research monitoring includes three types of projects as shown in Table 5-1 (pages 134 and 135): energy end-use monitoring, technology assessment, and diagnostic measurement (Misuriello 1987).

Forces Affecting the Development of Field Monitoring

It is difficult to understand and appreciate the state of the art of energy performance monitoring in buildings without some overall concept of the breadth and depth of this diverse field and some understanding of the forces that have influenced its development over the last 15 years. Indeed, one theme of this review is that load-research monitoring and issues-research monitoring are unevenly developed. Analysis of these differences can highlight major problem areas and point out directions for improvement. In addition, contrasts between the current state of development of these two families of monitoring projects have important implications for the building energy efficiency community.

In November of 1978, Congress enacted a five-part national energy program: National Energy Conservation Policy Act, Natural Gas Policy Act of 1978, Energy Tax Act of 1978, Power Plant and Industrial Fuel Use Act of 1978, and Public Utility Regulatory Policies Act of 1978 (PURPA). This wave of legislation, in addition to previous legislation (for example, the Energy Conservation Standards for New Buildings Act of 1976), established numerous energy

Table 5-1. Characteristics of major monitoring project types.

	Load Research		Issues Research	
	Aggregate Energy Use	Energy End Use	Technology Assessment	Diagnostics
Goals & Objectives	Infer average performance of large sample of buildings.	Determine characteristics of specific energy end uses within a building.	Measure field performance of building system technology, or retrofit measure in individual buildings.	Solve problems. Measure physical or operating parameters that affect energy use, or that are needed to model building or system performance.
Experimental Approach	Large, statistically designed sample with small number of data points per building. Often uses whole building utility meters.	Often use large statistically designed sample. Monitor energy demand or use profile of each end use of interest within building.	Characterize individual building or technology, occupant, and operation. Account and correct for variations.	Typically use one-time and/or short-term measurement with special methods such as infrared imaging[a], flue-gas analysis[b], blower door[c], and coheating[d].
Level of Montoring Detail	Average performance of sample, not individual building. Cannot explain variations between buildings. Often uses seasonal or annual data.	Detailed data on end uses metered. Collect building and operating data that affect end use.	Use detailed audit, submetering, indoor temperature, on-site weather, and occupant surveys. May use weekly, hourly, or short-term data.	Focused on specific building component or system. Amount and frequency of data varies widely between projects.
Examples	Class-load studies, cost-of-service studies, and load forecasting. Conservation or demand-side management program evaluation. Rate design.	Load forecasting. Identify energy conservation opportunities. Rate design.	Technology evaluation. Retrofit performance. Validate models and predictions.	Energy audit. Identify and solve O&M, IAQ, or system problems. Provide input for models. House doctor and building commissioning.

(continued on next page)

Table 5-1 ***(continued).***

a. Infrared imaging uses special devices to display and record the surface temperature of materials as measured from the infrared emissions of the surface. The device may also indicate the temperature differences across a surface, relative to a reference point. As an example, this technique can be used to assess effectiveness of wall insulation, since well-insulated portions of walls will appear "cooler" than poorly insulated sections, which appear "hot" (that is, allowing more conductive heat loss). Aerial infrared imaging has also been used to characterize the effectiveness of roof insulation in houses.

b. Flue-gas analysis measures parameters related to fossil-fuel combustion efficiency in boilers, furnaces, and other combustion devices. Flue-gas measurements include fuel-air ratio; oxygen, carbon dioxide, and carbon monoxide content of flue gas; and temperature of flue gas.

c. A blower door is a portable device containing a fan used to pressurize (or depressurize) a house in order to characterize air infiltration. The blower door is usually mounted temporarily in an exterior doorway. The blower door is calibrated so that the resulting pressurization readings can be converted into meaningful units such as effective leakage area or air changes per hour.

d. Coheating is a method used to characterize a building heat loss coefficient, usually in residences. Typically, portable electric heaters are used to raise the internal temperature of a house to a certain point, allowing the amount of required electric heat to be accurately measured. A coheating test may include observation of how long it takes for the house temperature to drop by a certain amount. Coheating tests generally employ statistical and engineering analysis of indoor and outdoor conditions.

Source: ASHRAE 1990.

policies and programs (among others, the Residential Conservation Service, the Institutional Conservation Program, weatherization programs, and Building Energy Performance Standards). This legislation also established an agenda of challenging research issues and, indirectly, a state and national research infrastructure to pursue them. Both load research and issues-research monitoring received their impetus from these laws.

Although utility load research was undertaken before the advent of PURPA, this legislation is generally regarded as a significant demarcation point in its development. PURPA required utilities to collect, analyze, and report on the energy usage patterns of their customers (AEIC 1987). With the advent of PURPA, utilities had to develop internal staff capabilities to perform class-load and cost-of-service studies.

Likewise, issues research monitoring was driven by the legislatively mandated need to provide data for program design and policy making. Users and providers of these data included many groups besides utilities. Since the 1970s, these two monitoring project families have developed much differently, as summarized in Table 5-2 (page 139). This table compares the two families according to a number of indicators of their strengths and weaknesses and of their general levels of development. Overall, load research seems more highly developed as a specialized discipline for the following reasons:

1. The scope of load research is comparatively narrow, allowing routine (but challenging) data collection and analyses. Capabilities can expand, using lessons learned from less complex experiments. Similar tasks are performed year after year by a large number of practitioners throughout a single industry with similar requirements across utilities.

2. Load research is vital for utility operation, primarily for regulatory requirements and ratemaking purposes. Load research can also identify the most profitable customers and end uses. As a result, load research receives internal management support and substantial funding.

3. Load research practitioners are organized through the Load Research Committee of the Association of Edison Illuminating Companies (AEIC). In addition to annually sponsoring national and regional conferences, AEIC has a professional development program. Between 1968 and 1987, more than 900 attendees participated in its Fundamentals of Load Research seminar (AEIC 1987). AEIC maintains a publications list of conference proceedings,

methodology guidelines, and a cumulative subject index for 1959–1984.

4. Load research benefits from the availability of specialized and reliable load survey recorder hardware designed to meet the data collection needs of practitioners. Due to the similarity of load research activities among utilities and to the continuing data needs of regulators, a competitive equipment market has developed.

In summary, load research is a strong and highly focused discipline that has both regulatory and corporate support, and, more importantly, commensurate funding. Its practitioners have also organized to provide themselves with the continuity of experience needed to carry out their mission.

In contrast to load research, monitoring building energy efficiency for the purpose of issues research is a discipline too broad to have yet found its boundaries. This is probably because issues research data have much different constituencies and purposes than load research data have. For example, issues research data often support a broad range of advocacy positions within the equally broad assemblage of organizations and individuals comprising the building energy efficiency and research community. (As used in this chapter, "advocacy" refers to active support for policies and programs that promote building energy efficiency, energy conservation, and DSM.)

Thus, in the last 15 years, issues research and its field monitoring component have lacked an enduring focus on a single topic. Consider the past ebb and flow in interest and importance accorded to field monitoring with respect to the following building energy efficiency topics:

- Building energy simulation model development and validation
- Infiltration, ventilation, and indoor air quality
- Active and passive solar systems (residential and commercial)
- Appliance efficiency standards
- Energy codes and standards for new residential and commercial buildings
- Behavioral effects
- Weatherization program measures
- Building energy efficiency labeling and rating systems
- Program evaluation

Much of this uneven interest in certain issues is due to the nature of energy-efficiency advocacy. Issues often rise to temporary prominence when competing interests are campaigning to have their points of view adopted by policy makers. Thus, data needs are sometimes temporal. This effect has had mixed impacts on issues research monitoring practitioners. As indicated in Table 5-2, for example, the broad scope and shifting emphasis of building energy efficiency issues does not facilitate the development of monitoring capability in any particular area. However, the shifting focus on issues (reflected quickly in the market for monitoring services) develops flexibility and creativity among practitioners, and these traits are probably the biggest strengths that issues research monitoring practitioners have. Organizational shortcomings in this field (such as lack of professional associations, training programs, and accessible literature) will probably be resolved as DSM and least-cost utility planning (LCUP) concepts impose some stability by establishing which energy-efficiency issues will need to be investigated on a continuing basis.

In many respects, the comparative states of development of these two families of field monitoring mirrors the current status of the building-energy-efficiency community. The years of advocacy for energy efficiency are paying dividends in the 1990s as formerly unorthodox points of view become mainstream and institutionalized through utility DSM and LCUP programs. Moreover, regulations requiring collection of field data for planning, evaluation, and utility performance incentives may well provide issues-research monitoring with the institutional stature and support it needs.

State of the Art of Field Monitoring for Issues Research

Examining the status of issues-research field monitoring in 1990 requires a variety of perspectives due to the diversity of field applications, practitioners, and end users of the information. Nonetheless, there are three major criteria by which to assess the diverse field monitoring efforts. These are:

1. Monitoring protocols. Monitoring protocols define the methodological aspects of field monitoring projects. The use of monitoring protocols generally indicates that all critical aspects of a monitoring project have been planned and specified in sufficient detail to yield useful project output. While the need for planning is seemingly obvious, shortcomings in project planning have been a persistent problem in certain types of field monitoring projects.

Table 5-2. Comparison of load research to issues research monitoring.

	Load Research	Issues Research
Impetus	Mid-1970s energy legislation.	Same.
Scope	Relatively narrow. Class load; cost of service; appliance studies; and some end-use profiles.	Very broad. All manner of residential, commercial, and multifamily sector issues.
Importance to Decision Makers	Data products vital to utility decision makers. Provides feedback on customer and end-use profitability. Data needed for rate making.	Uneven importance to diverse set of decision makers. Empirical data are often just one of many factors considered.
Continuing Focus	Long-term focus on basic scope of continuing interest to utility business operations and regulatory information requirements.	Often short-term focus on key policy issues where interest and importance may vary over time.
Development Characteristics	Expansion of capabilities based on refinement of previous experience.	Monitoring capabilities developed on one issue often not transferred to other issues. Experience often not capitalized on.
Practioners' Organization	Load Research Committee of the Association of Edison Illuminating Companies. National and regional chapter. Annual meetings.	No special-interest organization. Only three national conferences (1985, 1989, and 1990) to date.
Training	Strong AEIC training program for 20 years.	No organized training program.
Literature	Well-organized topical literature from 1959 to 1989.	No comprehensive bibliography or list of studies. Literature not generally accessible.
Methodology	Variety of published methodology handbooks.	Some methodology literature in last three years.
Hardware	Competitive market for specialized load survey recorders.	Wide variety of equipment, but few integrated systems. Special FDAS equipment developed in last five years.

2. Field data acquisition system (FDAS) technology. All field monitoring projects require special hardware to collect building energy performance data. Although composed of discrete components—sensors, data loggers, and supporting peripherals—the hardware must be viewed as a system. Accuracy and precision, the key indicators of data quality, can only be properly determined on a total system basis. Development of FDAS technology is primarily concerned with increasing its reliability and with the suitability of specific technology to specific applications.
3. Quality assurance and control (QA/QC). The data ultimately produced from a field monitoring project may have been subjected to numerous opportunities for error and measurement failure. Raw data typically come from a particular sensor or transducer and end up in a data table or chart. Hard experience has forced both private sector practitioners and research organizations to develop comprehensive QA/QC procedures to counter threats to quality in each element of the data path, or pipeline. Thus, accuracy and data capture rate (the percentage of possible data that could have been collected) are key indicators of the technology's level of development.

In short, a state-of-the-art field monitoring project will be one that has been comprehensively planned to yield specific, needed data products; utilizes appropriate and reliable FDAS equipment; and employs effective QA/QC procedures. In the final analysis, the state of the art of field monitoring is determined by how well the final data products answer the sponsor's research questions.

Monitoring Protocols

In recent years, field monitoring of the energy performance of buildings and appliances has been developing into a specialized discipline in its own right. This trend was particularly evidenced in the successful convening of the first National Workshop on Field Data Acquisition for Building and Equipment Energy Use Monitoring at Dallas, Texas, in October 1985 (ORNL 1986). Since then, two more national conferences (Fleming 1989 and 1990) and an international workshop (IEA1990) have been held on the subject of field monitoring.

As a developing discipline, field performance monitoring has been subject to introspection and self-criticism by its practitioners. Recurring themes have been the need for detailed planning of monitoring projects and the importance of establishing the objectives of the monitoring project. Monitoring protocols—standardized project planning methodologies—can assure that these concerns are identified and resolved prior to commencing the actual field monitoring activities.

A procedure for uniform development and documentation of monitoring protocols has been proposed (Misuriello 1987); it includes the following major elements:

1. A classification system that broadly groups various types of monitoring projects and associated protocols according to similarities of goals, general approach, and data requirements. Such projects and protocols would include aggregate energy use, energy end use, technology assessment, and building diagnostics.
2. A guide specification that provides a consistent format for communicating the methodological requirements of particular monitoring projects, including procedures for relating the project goals and objectives to the monitoring and research questions to be addressed; specifying data products, data analysis procedures and algorithms, and data points to be monitored; special concerns of hardware selection; and procedures for quality control.
3. Standard terminology and definitions that further specify the monitoring protocol requirements, including definitions of quantities to be measured (such as energy end uses) and published, referenced standards of measuring and testing.

Published Monitoring Protocols

The U.S. literature on building performance monitoring generally consists of case-study reports emphasizing data and findings for particular projects. Few references specifically treat the methodological aspects of monitoring projects.[1] With the recent attention to methodological issues by monitoring practitioners, more technical papers on this topic are beginning to appear. In addition, the American Society of Heating, Refrigerating and Air-Conditioning Engineers (ASHRAE) plans to include a new chapter on monitoring protocols in the 1991 ASHRAE Handbook (ASHRAE 1990). This new chapter will emphasize monitoring project planning methodology, as shown in Figure 5-1 (next page). Other organizations, such as the American Society for Testing of Materials (ASTM), plan to issue standards for energy monitoring in residences.

Recently published monitoring protocols that generally adhere to the practice recommended previously include the following:

---

1. Early examples of monitoring protocols include the Solar Energy Research Institute Residential Class B Passive Solar Performance Monitoring Program (Swisher et al. 1982) and the Monitoring Handbook for Residential HVAC Systems (EPRI 1983).

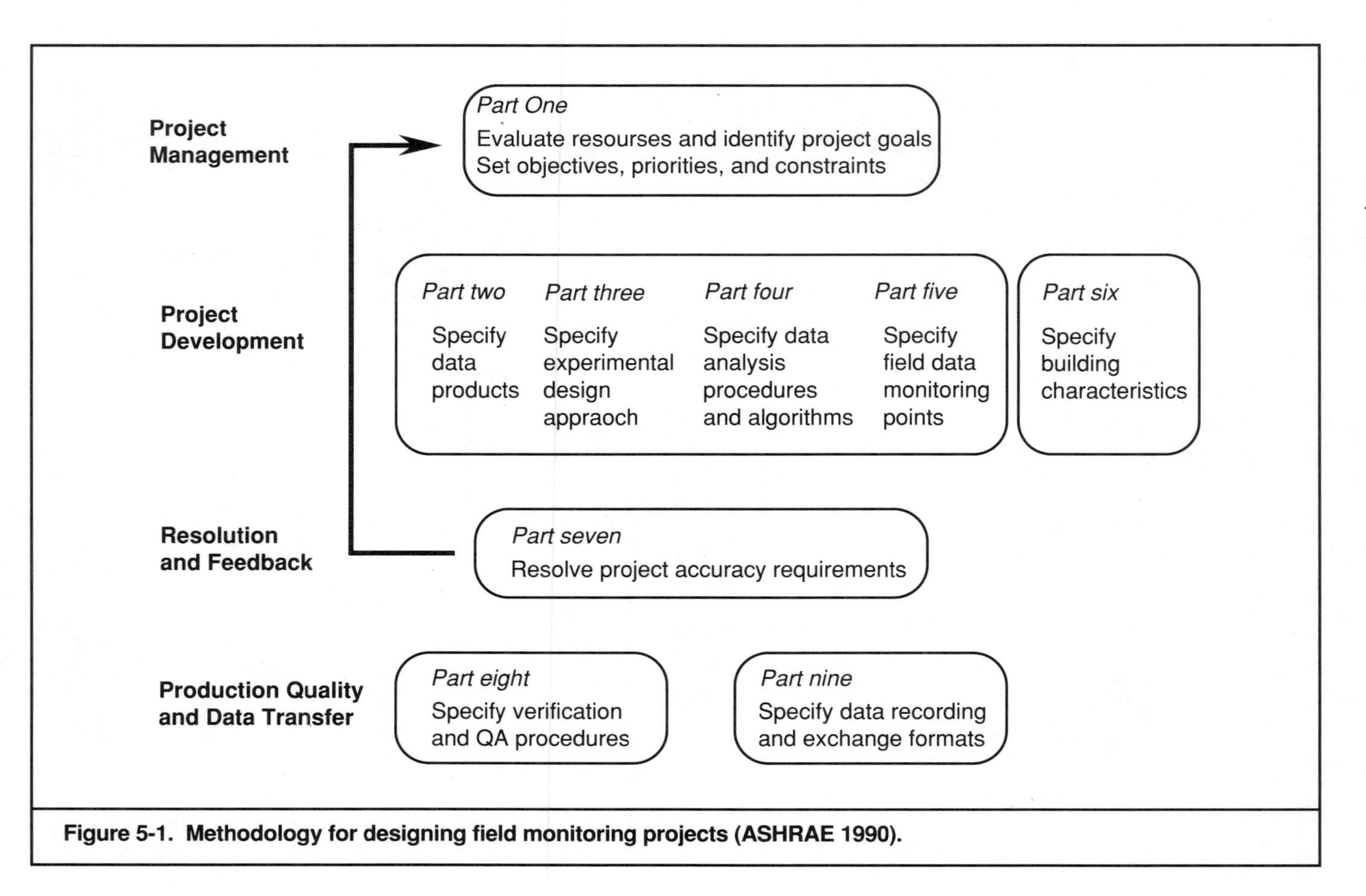

Figure 5-1. Methodology for designing field monitoring projects (ASHRAE 1990).

1. A results-oriented methodology for monitoring HVAC equipment in the field (Hughes et al. 1987). Based on project experience at over 140 residential and commercial building sites, this protocol has been designed to yield reliable and accurate field data through close attention to quality control. It is intended to provide data for developing and verifying HVAC design and prediction models for technology assessment purposes. Its major application is for the assessment and subsequent commercialization and technology transfer of new or advanced technologies. The major contributions of this protocol include strong emphasis on project planning and quality control, advanced documentation and communications features through a uniform data acquisition and analysis (DAA) plan, capability to collect nonmetered data critical to meeting project goals and objectives, and procedures for the staffing and organization of technology assessment projects
2. A commercial-building energy end-use monitoring protocol for utility load research (Mazzucchi 1987). This protocol was developed to collect data for the End Use Load and Consumer Assessment Program (ELCAP), a large-scale (250 commercial building sites) monitoring effort conducted for the Bonneville Power Administration. The data collected using the ELCAP protocol will be used to predict future electrical demands in the region and to develop methods to assess and acquire cost-effective building conservation resources. The unique aspects of this protocol include a commercial-sector sampling plan that addresses three strata of building sizes and ten building types, detailed procedures for developing site-specific measurement plans, and procedures to collect detailed building characteristics data. A significant contribution of the ELCAP measurement plan has been in the area of standard terminology and definitions, defining 20 specific energy end-use categories and identifying codes for commercial building equipment. This approach greatly increases data collection consistency, since data definition errors are minimized.
3. Data specification guidelines to determine actual energy conservation retrofit savings in single-family residences (Ternes 1987) and multifamily residences (Szydlowski and Diamond 1989). Developed by the U.S. Department of Energy (DOE), these protocols recommend a before-and-after experimental design approach and specify a minimum data set to be collected to determine actual retrofit performance on a normalized basis

(Fracastero and Lyberg 1983). The basic data set includes time-series data such as end-use quantities, consumption by fuel type, and indoor temperature. Data are also collected on a one-time basis for descriptive information, occupant interviews, infiltration measurements, and assessments of retrofit installation quality. These protocols have standardized the experimental design and data collection specifications so that independent researchers can obtain comparable results.

4. A protocol for monitoring energy efficiency improvements in commercial buildings (MacDonald, Sharp, and Gettings 1989). This protocol has been developed by DOE for use in field monitoring studies of energy improvements to commercial buildings. It is intended to support the development of long-term energy performance tracking methods, with building descriptive information supporting comparisons of efficiency between buildings. The strategy used for this protocol is to specify data requirements in four general categories:
   - Project or program description
   - Analysis methods and results: summary of analysis methods, experimental design, and project results
   - Performance data: summary of monthly billing data, submetered or detailed energy consumptions, inclusion of demand data, and temperature and weather data
   - Building core data: survey data that describe each building and associated building systems, functional use areas, tenants, schedules, base energy data, and energy improvements

   Since data are expected to be collected by different investigators, the protocol helps to ensure that data from projects are reasonably consistent and comparable.

The four example protocols described above include all elements of an acceptable monitoring protocol: a project-type classification, a guide specification for project activities, and standard terminology and definitions. While each has been developed for a specific purpose, all serve as examples of good monitoring practice, and can be utilized or adapted by other researchers with similar data and analytical needs.

Field Data Acquisition System Technology

FDAS technology topics most important to field monitoring are equipment reliability, technical advances that improve data collection and allow researchers to carry out tasks they could not accomplish before, and limitations imposed by the current technology.

Equipment Reliability as a Development Force

Until recent years, the unreliability of field equipment had been one of the most significant problems faced by practitioners. Fifteen years ago, for example, the DOE active-solar program was routinely plagued by FDAS failures, resulting in very low data capture. However, certain FDAS equipment today has reached the point that contracts for monitoring projects often contain performance standards for data capture: in some instances, failure to collect less than 95% of available monthly data can reduce a monitoring contractor's fee. In this regard, probably the single most important FDAS technical advance was the development of the packaged FDAS, which generally has the following characteristics:

- System components housed in a single rugged field enclosure
- Single-board microprocessor configuration with substantial memory for data storage
- Supporting electronic components (signal conditioning, analog-to-digital conversion, and so forth) installed on the microprocessor board
- Battery backup power supply
- Passive heat dissipation (no cooling fan)
- On-board "firmware" (software in nonvolatile memory)
- Integral telephone modem for data retrieval
- Measurement transducers installed within enclosure, or integral with microprocessor board
- On-board terminals for field connections
- Communications interface for site configuration setup and diagnostics
- Sophisticated programmability, such as software branching often required in advanced technology assessment projects
- Daisy-chain capability to link multiple FDAS units at a single monitoring site

As is evident from this list, the packaged FDAS provides solutions to many practical field operation difficulties. For example, housing all or most components in a single enclosure protects the equipment from field hazards. Elimination of a cooling fan (by using passive heat dissipation) prevents overheating of components when air particle filters become clogged. Substantial integration of electronic components and on-board firmware also increases reliability. This integration also greatly simplifies field installation since sensor leads are simply wired to the unit.

Packaged FDASs also largely eliminate custom system engineering by the end users. Prior to packaged FDASs, field monitoring practitioners had to assemble systems from components obtained from many different vendors. Practitioners were responsible for all design and specification work and for assuring the physical and electronic compatibility of all components as a system, in addition to programming the data loggers. Modern packaged FDASs, therefore, not only reduce possible design and compatibility errors, but also reduce engineering time and costs.

The Impact of the Large-Scale Multisite Project

Prior to the advent of packaged FDAS, practitioners experimented with the use of personal computers (PC) as the core of the FDAS. Special plug-in data acquisition boards facilitated the interfacing of sensors to the PC. The PC itself had many of the functional attributes of the modern FDAS: programmability, on-board data storage via floppy disk or tape drive, and modem capabilities. The PC also had the ability to auto-start, or "boot," after a power failure. More importantly, the PC offered the advantage of being capable of performing data analysis with the same hardware used for data collection.

These PC-based FDASs were predominantly used on single-site demonstration and technology assessment sites, where project staff could regularly visit to check operation and perform system maintenance. Thus, the monitoring site was in many instances a field research station. However, in the mid-1980s, the nature of field monitoring projects began to change, and this FDAS approach was no longer practical. Project sponsors began to commission large-scale multisite field monitoring projects. These projects were mostly oriented toward residential- and commercial-sector end-use load data collection, but some significant multisite technology assessment programs were initiated as well. This new level of program monitoring had the following significant impacts on field monitoring practice:

- Widespread use of the packaged FDAS for reliability and consistency of installation.
- Formal use of documentation to communicate procedures to project participants, including standardized definitions of end uses

and their associated equipment. On larger programs, documentation included special manuals and guidelines for preparing site measurement plans, installing FDAS equipment, and performing building characteristics audits.

- Measurement device innovation, especially for measurement of electric power (Schuster and Tomich 1986). In particular, low-cost multipoint digital watt-hour meters on a single circuit board were developed.
- Development of specialized organizations with professional and trade skills to handle multidisciplinary requirements of large-scale multisite programs and the new technologies they employ. Note that these organizations developed within private consulting firms and national laboratories, and generally not within electric utilities.
- Establishment of specialized and sophisticated central data collection and processing facilities. Large-scale multisite projects required powerful computing facilities to collect data from many geographically remote sites, perform quality control checking, maintain data base and reporting facilities, and facilitate data analysis (see Figure 5-2, next page).

In summary, the state of the art of field monitoring equipment and procedures has mostly improved over the past 15 years as the character of monitoring projects has evolved. The advent of large-scale multisite projects (mostly for end-use monitoring) accelerated this change, forcing practitioners and equipment manufacturers to create new FDAS equipment to accommodate increasingly complex client information needs. The outcome of this evolution has been the establishment of a reliable FDAS platform readily adaptable to future programmatic needs, with a relatively low marginal cost for adding monitoring sites to the system.

Gaps in FDAS Technology

Even though the quality and technique of field monitoring continues to improve, there are a few significant gaps in FDAS technical capabilities. These technical gaps are mostly related to a more detailed understanding of building energy performance, but at least one concerns basic collection of consumption data for natural gas end uses. These gaps are as follows:

1. Natural gas end-use consumption. Virtually all large end-use studies to date have focused on electricity. However, policy and regulatory attention is also turning toward natural gas, and detailed end-use information may be required in the future to

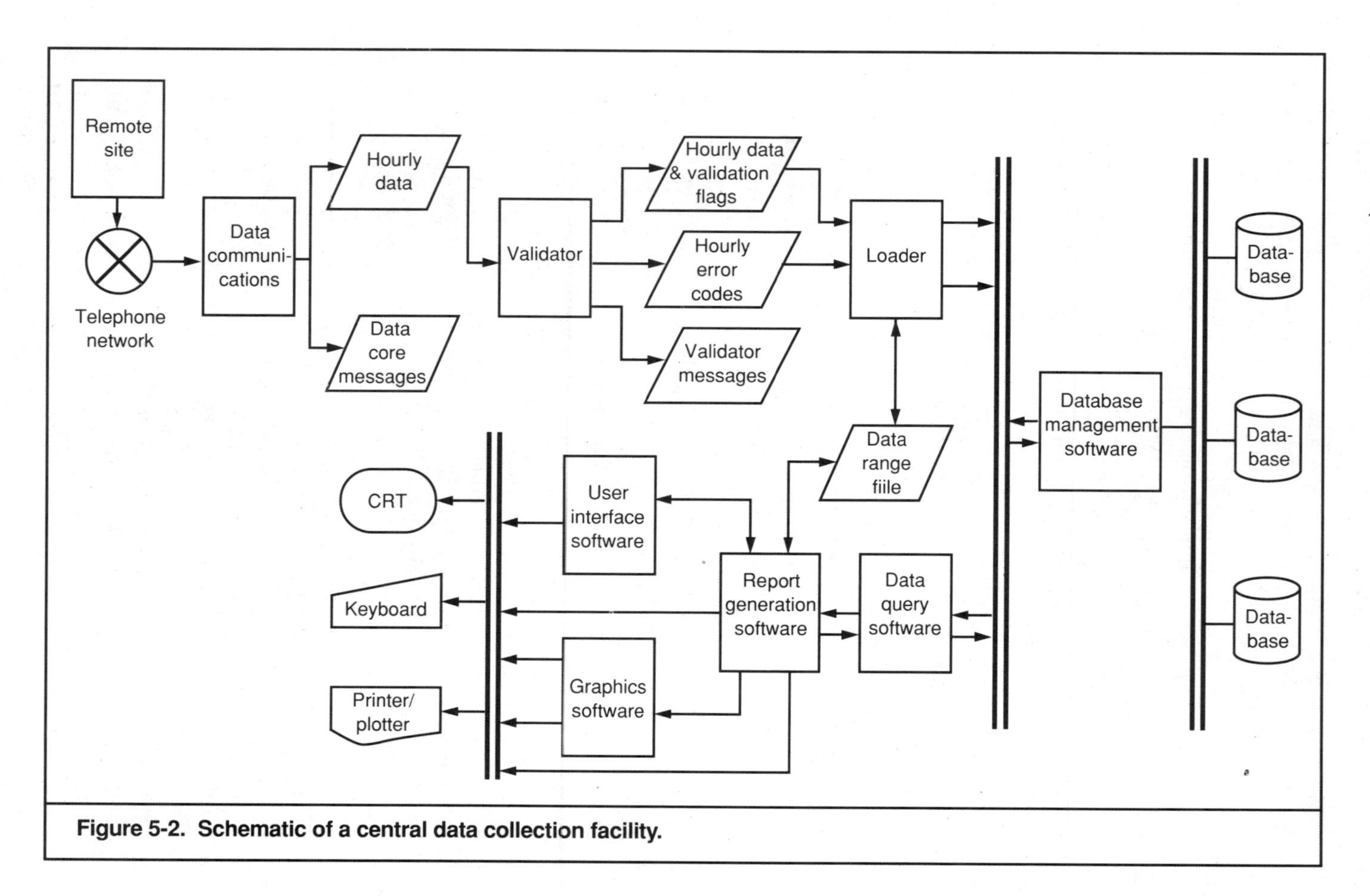

Figure 5-2. Schematic of a central data collection facility.

support DSM and LCUP programs. One strong barrier to gas end-use data collection is the significant cost of installing additional standard gas meters (with pulse signal output adapters) for each monitored end use, along with required shut-off values and bypass piping. However, little or no research and development has been performed to develop products that significantly reduce monitoring costs with innovative metering devices as was done for the large-scale electric studies (ELCAP). One possible approach is to develop reliable procedures that correlate gas end-use equipment run time to consumption, since gas is usually burned at a fixed rate in many devices (furnaces, boilers, water heaters, dryers). Some research groups have used photoelectric sensors, flue temperature sensors, and control signals from equipment thermostats to monitor run time. Variable-rate appliances, such as those for cooking, may require a different approach.

2. System output measurement. Most end-use studies, and some technology assessment projects, only measure energy consumption, and not the useful output of systems and equipment. Thus, in-situ energy efficiency is not determined. Continuous, long-term collection of field data on cooling coefficient of performance, seasonal energy efficiency ratio, annual fuel utilization efficiency, and water heater energy factors is expensive and sometimes difficult. Only a limited amount of work has been done in this area (Szydlowski and Cleary 1988).

3. Thermostat setting monitoring. Considering the interest in behavioral issues and phenomena such as the "take-back effect," it is surprising that a simple, low-cost thermostat setting monitor is not commercially available. Such a device has been demonstrated on research projects (Harrje, Gadsby, and Grot 1986). Substantial analytical effort is applied to discerning this behavior through statistical means and occupant surveys, yet hardware approaches to directly monitor this parameter are overlooked.

Quality Assurance and Control

QA/QC has always been of concern to field monitoring practitioners. The range of data quality problems that practitioners encounter in the field is astonishing, yet much of the published information has been anecdotal (DeCicco et al. 1986). The advent of large-scale, multisite monitoring projects has helped fill this literature gap. For example, many of the published papers concerning the BPA's

ELCAP program have addressed quality control (Pearson, Stokes, and Crowder 1986; Halverson, Conner, and Caplinger 1988; and Stoops 1989). More recently, practitioners' efforts to develop monitoring project planning protocols recognize that QA/QC procedures must be an integral part of the planning process (ASHRAE 1990).

Quality control on field monitoring projects must be comprehensive and, therefore, must involve FDAS hardware, data processing software, and persistent project management procedures. Unlike FDAS hardware, QA/QC cannot be acquired in packaged form. Therefore, the quality of such procedures is embodied in, and distributed throughout, all project resources, including personnel. Several key points in this regard are made by Pearson, Stokes, and Crowder (1986):

- For any large-scale metering project, data verification must be highly automated; failure to automate will lead to enormous demands on staff time.
- The single best assurance of data quality is the execution of redundant measurements. Implementation of redundant measurements provides powerful diagnostic capabilities.
- Installation of metering hardware is a complex task, and errors should be expected. Real-time tests during installation substantially reduce the frequency of many common installation errors but do not eliminate them.
- There are many system components that can fail, and verification procedures must be developed to detect failures in any of them.
- Even with substantial automation in data verification and careful development of installation procedures, data verification is a time-consuming and expensive task for which provisions must be made in project plans and budgets.

These are useful points in determining what must be done; however, they must be integrated into each phase of the project by design, and must address FDAS hardware and the types of data collected. Table 5-3 summarizes how this goal may be accomplished, leading to the timely detection and correction of problems.

FDAS Quality Control

The increased reliability of a modern, packaged FDAS eliminates many, but not all, of the traditional hardware problems. It is still necessary to bench-test operation and performance of each system before installations. Some testing may be extensive. For example, some leading field monitoring organizations use electric load

Table 5-3. Quality assurance elements for field monitoring projects.

Time Frame	Hardware	Engineering Data	Characteristics Data
Start-up	Bench Calibration[a] Field Calibration[a] Installation verification [a]	Installation verification[a] Collection verification[a,b] Processing verification[a,b] Result production[a,b]	Field verification[a] Completeness check[a] Reasonableness check[a,b] Result production[a,b]
Ongoing	Functional testing[a] Failure mode diagnosis[c] Repair/maintenance[d] Control change[a]	Quality checking[b] Reasonableness checking[b] Failure mode diagnosis[c] Data reconstruction[d] Control change[a]	Problem diagnosis[c] Data reconstruction[d] Control change[a]
Periodic	Preventive maintenance[a] Calibration[a]	Summary report preparation and review[b]	Scheduled updates resurvey[a] Summary report preparation and review[b]

a. Actions to ensure good data
b. Actions to check data quality
c. Actions to diagnose problems
d. Actions to repair problems

Source: ASHRAE 1990

simulators to bench-test performance of FDAS kilowatt-hour transducers. Once installed, the complete FDAS is extensively calibrated and tested, often using portable computers to exercise and diagnose system operation against an extensive checklist of procedures. Reliable communications with the central data collection facility are also tested. Ongoing and periodic maintenance include correcting any deficiencies or failures identified through data checking, and performing preventive maintenance on system components.

Calibration and periodic recalibration of FDAS sensors are also critical for data quality control (Haberl, Claridge, and Harrje 1990). Procedures for calibration of many types of sensors are available from organizations such as The National Institute of Standards and Technology (formerly the National Bureau of Standards), the American Society for Testing and Materials, and the Instrument Society of America. A thorough discussion of sensor applications and calibration is given by Hurley (1986).

Verification of Equipment Being Monitored

Monitoring teams must also verify that the equipment components to be analyzed are performing as intended. For example, if heat pump performance is being monitored, it is important to verify that the heat pump itself is operating correctly. Many parameters can be checked to verify proper equipment performance. A manufacturer's data provide various pressures, temperatures, and flows throughout the equipment, as well as power consumption, capacities, and performance measures. Although one should not expect performance data to match manufacturer data exactly, performance data should fall within the published range.

Data Quality Control

Data quality control has both hardware and data processing aspects and must begin before FDAS installation at a monitoring site. The initial QA/QC action is the preparation of a detailed site measurement plan that specifies complete details on which data points are to be taken. This plan is essentially a design and installation procedure to be implemented by field installers, and it facilitates subsequent data checking and analysis (Mazzucchi 1987, Hughes and Clark 1986). In terms of data QA/QC, a measurement plan provides the following:

- Parameters to be measured are correctly defined and classified, and individual electrical circuits are assigned to end-use categories.
- Critical data points are identified for redundant measurement to protect against measurement failure; for example, duplicate temperature sensors may be imbedded in floor or ceiling slabs where replacement is not practical.
- Electric panel sum checking is used, primarily on end-use metering projects, to assure that individual electric circuit measurements add up to total panel loads.

A good measurement plan and rigorous FDAS installation verification are necessary and effective, but the process must be continued by subjecting collected data to a variety of quality control checks. Most checks are automated through central data collection facility software. Successful implementation, however, requires that data be retrieved frequently (daily if possible) so that problems are identified and corrected quickly. Some of the most frequently used data-checking procedures are as follows:

1. Upper and lower bounds checking. The range of various temperatures, flows, pressures, consumptions, run times, and so

forth, can be predicted before a site is instrumented, and adjusted further after data collection has started. Data points falling outside of an assigned range during automatic incoming data verification should be flagged and placed in an error detection report. If redundant data points are used for critical measurements, the range check should be augmented to compare data from these backup sensors.

2. Maximum rate-of-change checking. To take one example, the sudden grounding or ungrounding of a sensor, such as a thermocouple, may change the magnitude of its output. Although the output may remain within expected limits, the rate of change between recording intervals may be beyond expected limits.
3. Relational data checking. A relational check involves examining the relational logic of the output of one data point to another. This verification step is extremely useful for detecting equipment failures, which may go undetected in a range check. For example, if fan energy use is detected while no air flow is reported, an error would be noted.
4. Comparison of calculated parameters to expected values. Calculated variables can often be compared to expected values to determine whether systems are performing as anticipated. This comparison might include such parameters as air temperature drop across a cooling coil, compressor or fan electrical consumption, Btu output, system efficiency, or COP versus outside air temperature. These types of checks are usually applied to steady state conditions or average hourly values rather than to transient conditions.

In summary, field monitoring projects cannot provide reliable data of known quality without the type of QA/QC procedures described above. Advancements in this area are expressed as organizational capabilities rather than as products. In addition, field monitoring QA/QC is often specific to a particular project and to different categories of projects. For example, end-use monitoring QA/QC procedures can be different from those for technology assessment projects. Finally, information on this subject is not widely available or readily transferrable to those entering the field.

Future Directions for Issues-Research Field Monitoring

Field monitoring has always played an important, but supporting, role in the advocacy of energy efficiency. As discussed earlier, its specific applications have changed focus over time due to the changing perspectives and imperatives of its sponsors. While this instability is likely to continue, many policies supporting increased energy efficiency are being adopted and institutionalized, and this trend should thereby bring some stability and regularity to certain monitoring applications. Thus, speculation on the future of field monitoring requires some guesswork about the future of the energy-efficiency movement.

It seems probable that the growing adoption of DSM strategies by utilities and the expanding interest in LCUP by regulators will require increased use of field-monitored data. Some areas of field monitoring may become more like load research in terms of their role in the operation of utilities and in the regulatory process. In this context, likely applications for field monitoring include the following:

1. LCUP applications. For DSM evaluation, regulators need many of the same field data that utilities need. However, the data collected will be used to monitor utility performance toward DSM goals and provide input to prudency reviews. Regulators will be interested in end-use load information with which to establish energy consumption baselines; they will also need data on the performance of DSM measures in both full-scale DSM programs and pilot program efforts. Field-monitored DSM performance data may also be used to determine utility DSM performance incentives.
2. DSM program implementation. Field monitoring techniques may also find application in the implementation of pilot and full-scale DSM programs. For example, short-term diagnostic field measurements can identify sources of energy waste and identify specific retrofit opportunities in commercial buildings (Misuriello 1988; Haberl and Vajda 1988). Similarly, house doctor techniques can be applied to residences for focusing conservation programs. Utilities are greatly interested in energy-efficient design programs for new construction, and field measurement could become a step in the commissioning of buildings. Other potential uses of field monitoring techniques for DSM implementation could be in the area of innovative rate design. For example, utilities may find it advantageous to experiment with special end-use tariffs. Such an approach will require monitoring protocols in addition to end-use metering.
3. DSM program evaluation. Regulatory pressure and the desire for profitability will probably induce utilities to determine actual

effects from DSM programs, and to determine if the programs are truly cost-effective. Such efforts are now under way in New England and the Pacific Northwest. This activity is not so much research-oriented as it is geared to providing feedback on utility operations. Load research techniques and equipment (in their present form) probably cannot provide enough detail for this purpose.

Improvements in Monitoring Protocols

Even though field monitoring practitioners are skilled, experienced, and resourceful, they are not well enough organized as a profession to document methodology or to develop standard procedures. The issues research case study orientation of past monitoring projects, coupled with low standing of DSM staff within utilities, has made standardization impractical. Although some monitoring protocol development has taken place, more work is needed to prepare the profession for its potentially significant expansion in the DSM and LCUP areas outlined above. In that role, field monitoring practitioners will need to become more like their load research colleagues in terms of professional organization, procedure documentation, and training.

The development of protocols should focus first on key program evaluation areas, such as lighting and HVAC system retrofits. New techniques in these areas should incorporate a standard data analysis procedure that identifies the retrofit effect, normalizes for external factors, such as weather and occupancy, and provides for year-to-year tracking of energy savings. These methods also need to address the problem of building size. For example, it is costly and impractical to instrument a 500,000-square-foot facility to completely measure all lighting panels for a retrofit project. Some type of reliable sampling procedure needs to be developed.

Monitoring methodology can also benefit from the introduction of expert systems. For example, the preparation of site measurement plans is, to a large extent, rule based and repetitive. These characteristics make preparation of measurement plans a potentially useful application for an expert system, the use of which could reduce the costs of field monitoring.

Current field monitoring methodology is oriented toward long-term observation (as opposed to testing under controlled conditions) of the energy performance of buildings. Some researchers have developed experimental methods for short-term testing of residential building systems to reduce the time and cost for obtaining useful results (Subbarao 1988; Duffy and Saunders 1987). These methods generally predict or extrapolate heating season thermal performance using simulation models and parameter inputs estimated from results of various one-time tests (coheating, infiltration, and so on). Thus, this type of monitoring protocol integrates a test procedure, an analysis method, and a field instrumentation package. These short-term methods have had limited application but promising results. With more development and application experience, short-term methods could be applied to home energy rating systems, before-and-after retrofit analysis, and development of energy usage baselines for planning and forecasting.

FDAS Hardware Advances

Future FDASs will likely continue to improve in capability and drop in cost. In the short term, the current configuration of the packaged FDAS will probably retain its basic characteristics with periodic refinements. However, in the mid- to long term, FDAS devices with completely new characteristics are likely to emerge.

In the mid-term, direct hardwire connections between sensors and FDAS units may be replaced with power-line carrier transducers or short-range radio frequency devices. Available technology such as the Electric Power Research Institute's Electric ARM is used mostly for residential load research, appliance studies, and some limited applications in small commercial buildings (EPRI 1988). However, this technology has not penetrated the large-scale end-use monitoring and technology assessment markets to a significant extent. With sufficient demand for rugged, reliable, and low-cost devices, this technology should find greater use.

Over the long term, nonintrusive load monitoring systems may prove to be the ultimate systems if the considerable technical challenges can be overcome. Nonintrusive monitoring systems do not directly measure each device; rather, they use a microprocessor-based analysis of power flow characteristics through the main meter to identify the pattern and energy use of individual appliances (EPRI 1988). Since they can be installed in a meter socket or at a utility pole, installation costs are greatly reduced. This device is now in the prototype testing stage, and is intended for load research applications.

Summary

This chapter has attempted to provide a comprehensive overview of the state of the art in field monitoring in 1990 and a retrospective view of the issues and forces that have shaped field monitoring.

The state of the art in this field cannot be expressed according to a single index. As discussed, there are three major elements to consider: monitoring protocols, FDAS hardware, and quality control procedures. In reviewing these elements, we find that this technical field is strong and versatile. However, it does lack many characteristics pertaining to a strong professional organization among practitioners. These shortcomings will probably be corrected as widespread adoption of DSM and LCUP require a higher level of performance from monitoring practitioners.

References

AEIC. See Association of Edison Illuminating Companies.

American Society of Heating, Refrigerating and Air-Conditioning Engineers. 1990. "Building Energy Monitoring." Chap. 37 in *1991 ASHRAE Handbook*. Atlanta: American Society of Heating, Refrigerating and Air-Conditioning Engineers. Photocopy of draft to be published in spring 1991.

ASHRAE. See American Society of Heating, Refrigerating and Air-Conditioning Engineers.

Association of Edison Illuminating Companies. 1987. *Report of the Load Research Committee 1986–1987*. New York: Association of Edison Illuminating Companies.

DeCicco, J., A. McGarity, L. Norford, and L. Ryan. 1986. "Instrumented Buildings: Experiences in Obtaining Accurate Data." In *Proceedings from the ACEEE 1986 Summer Study on Energy Efficiency in Buildings*. Vol. 9. Washington, D.C.: American Council for an Energy-Efficient Economy.

Duffy, J., and D. Saunders. 1987. *Low-Cost Methods for Evaluation of Space Conditioning Efficiency of Existing Homes*. Upper Marlboro, Md.: National Association of Homebuilders.

Electric Power Research Institute. 1983. *Monitoring Methodology Handbook for Residential HVAC Systems*. EM-3003. Palo Alto, Calif.: Electric Power Research Institute.

——. 1985. Survey of Residential End-Use Projects. EM-4578. Palo Alto, Calif.: Electric Power Research Institute.

——. 1988. Energy Utilization Catalog—Products from R&D. Palo Alto, Calif.: Electric Power Research Institute.

EPRI. See Electric Power Research Institute.

Fleming. 1989. *Proceedings of the End-Use Load Information and Application Conference*. Syracuse, N.Y.: The Fleming Group.

——. 1990. *Proceedings of the Conference on End-Use Load Information and Its Role in DSM*. Syracuse, N.Y.: The Fleming Group.

Fracastero, G., and M. Lyberg. 1983. *Guiding Principles Concerning Design of Experiments, Instruments, Instrumentation, and Measuring Techniques*. Stockholm: Swedish Council for Building Research.

Haberl, J., and J. Vajda. 1988. "Use of Metered Data Analysis to Improve Building Operation and Maintenance: Early Results from Two Federal Complexes." In *Proceedings of the 1988 ACEEE Summer Study on Energy Efficiency in Buildings*. Vol. 3. Washington, D.C.: American Council for an Energy-Efficient Economy.

Haberl, J., D. Claridge, and D. Harrje. 1990. "The Design of Field Experiments and Demonstrations." In *Proceedings of the IEA Field Monitoring Workshop*. Gothunburg, Sweden: International Energy Agency.

Halverson, M., C. Conner, and J. Caplinger. 1988. "Dealing with Data in Large End-Use Load Metering Projects—Data Quality and Data Access." In *Proceedings of the 1988 ACEEE Summer Study on Energy Efficiency in Buildings*. Vol. 10. Washington, D.C.: American Council for an Energy-Efficient Economy.

Harrje, D., K. Gadsby, and R. Grot. 1986. "Alternate Methods to Acquire Energy Usage and Occupancy Data." In *Proceedings of the National Workshop: Field Data Acquisition for Building and Equipment Energy-Use Monitoring*. ORNL CONF-8510218. Oak Ridge, Tenn.: Oak Ridge National Laboratory.

Heidell, J., R. Mazzucchi, and R. Reilly. 1985. *Commercial Building End-Use Metering Inventory*. PNL-5027. Richland, Wash.: Pacific Northwest Laboratory.

Hughes, P., and W. Clark. 1986. "Planning and Design of Field Data Acquisition and Analysis Projects: A Case Study." In *Proceedings of the National Workshop on Field Data Acquisition for Building and Equipment Energy-Use Monitoring*. Oak Ridge, Tenn.: Oak Ridge National Laboratory.

Hughes, P., R. Hough, R. Hackner, and W. Clark. 1987. "Results-Oriented Methodology for Monitoring HVAC Equipment in the Field." *ASHRAE Transactions* 93.

Hurley, W. 1986. "Measurement of Temperature, Humidity, and Fluid Flow." In *Proceedings of the National Workshop: Field Data Acquisition for Building and Equipment Energy-Use Monitoring*." ORNL CONF-8510218. Oak Ridge, Tenn.: Oak Ridge National Laboratory.

IEA. See International Energy Agency.

International Energy Agency. 1990. *Proceedings of the IEA Field Monitoring Workshop*. Gothunburg, Sweden: International Energy Agency.

MacDonald, J., T. Sharp, and M. Gettings. 1989. *A Protocol for Monitoring Energy Efficiency Improvements in Commercial and Related Buildings*. ORNL/CON-291. Oak Ridge, Tenn.: Oak Ridge National Laboratory.

Mazzucchi, R. 1987. "Commercial Building Energy Use Monitoring for Utility Load Research." *ASHRAE Transactions* 93.

Misuriello, H. 1987. "A Uniform Procedure for the Development and Dissemination of Monitoring Protocols." *ASHRAE Transactions* 93.

Misuriello, H. 1988. "Instrumentation Applications for Commercial Building Energy Audits." *ASHRAE Transactions* 95.

Oak Ridge National Laboratory. 1986. *Proceedings of the National*

Workshop on Field Data Acquisition for Building and Equipment Energy-Use Monitoring. Oak Ridge, Tenn.: Oak Ridge National Laboratory.

ORNL. See Oak Ridge National Laboratory.

Pearson, E., G. Stokes, and S. Crowder. 1986. "Data Verification in End-Use Metering." Commercial End-Use Metering Workshop. EM-4393. Palo Alto, Calif.: Electric Power Research Institute.

Schuster, G., and S. Tomich. 1986. "ELCAP Instrumentation." Commercial End-Use Metering Workshop. EPRI-EM-4393. Palo Alto, Calif.: Electric Power Research Institute.

Stoops, J. 1989. "Performance Monitoring Protocols." In *Proceedings of the End-Use Load Information and Application Conference*. Syracuse, N.Y.: The Fleming Group.

Subbarao, K. 1988. *PSTAR—Primary and Secondary Terms Analysis and Renormalization: A Unified Approach to Building Energy Simulations and Short-term Monitoring*. SERI/TR-254-3175. Golden, Colo.: Solar Energy Research Institute.

Swisher, J., K. Harr, D. Frey, and M. Holtz. 1982. "Performance Monitoring of Passive Solar Residences at the Class B Level." In *Proceedings of the 1982 ACEEE Summer Study on Energy Efficiency in Buildings*. Washington, D.C.: American Council for an Energy-Efficient Economy.

Szydlowski, R., and P. Cleary. 1988. "In-Situ Appliance Efficiency Audit Procedures." *ASHRAE Transactions* 95.

Szydlowski, R., and R. Diamond. 1989. *Data Specification Protocol for Multifamily Buildings*. LBL-27206. Berkeley, Calif.: Lawrence Berkeley Laboratory.

Ternes, M. 1987. "A Data Specification Guideline for DOE's Single-Family Building Energy Retrofit Research Program." *ASHRAE Transactions* 93.

Chapter 6

Low-Income Weatherization: Past, Present, and Future

Jeffrey Schlegel, *Wisconsin Energy Conservation Corporation*
John McBride and **Stephen Thomas,** *National Center for Appropriate Technology*
Paul Berkowitz, *Public Service Commission of Wisconsin*

Low-income weatherization programs are at a critical juncture. Major changes are occurring in funding sources, program delivery systems, and the level of technical sophistication. At the same time, the original purpose of the program, easing the burden of energy costs on low-income households, is as important as ever. Government and utility programs are struggling with how to deliver effective services designed to provide significant energy savings in an era of dwindling or changing resources.

This chapter summarizes the history of low-income weatherization programs from its roots as an emergency program operated by the U.S. Community Services Administration in response to the Arab oil embargo of 1973. Current program practices are described in a national survey of weatherization providers conducted by the National Center for Appropriate Technology (NCAT). Information is presented on the evolution of weatherization technology, the disappointing level of energy savings achieved by the program, the emergence of the utility industry as a funding source, client/customer education, the evolution of field personnel from estimator to auditor to technician, and the importance of ongoing program evaluation. Finally, emphasis is placed on the emerging trends and future directions of low-income weatherization based on the authors' own knowledge and discussions with the experts in the field.

Background: Energy and Low-Income Households

The United States is a nation of extremes—a land of poverty in the midst of plenty. In 1987, when the median income for a family of four was $36,800, over 14% of the United States population lived below the poverty level ($12,000 for a family of four). Perhaps more significantly, about 20% of the children in the United States live in families with incomes below the poverty level (Bane and Ellwood 1989). In addition, the economic recovery of the 1980s has bypassed the poor. While more people have become wealthy, the plight of the poor has not improved during the 1980s (Bane and Ellwood 1989).

In 1984, the average low-income American family spent 94% of its income on housing, food, and home energy (Economic Opportunity Research Institute 1986). Only 6% of family income remained for other needs such as medical expenses, transportation, and clothing. While low-income families consume 22% less energy and pay 25% less for utilities than the nonpoor, they use 20% more energy per square foot of living space (Vine and Reyes 1987). Low-income families also spend about 25% of their income for energy compared to 7% for the nonpoor (Vine and Reyes 1987).

Consumer expenditure surveys and other statistics reveal that low-income families, on average, have incomes significantly less than their expenses (U.S. Department of Commerce, Bureau of the Census 1988; U.S. Department of Labor, Bureau of Labor Statistics 1988). Hence, low-income families operate with deficit budgets and seek credit wherever they can. One common source of credit for low-income people is their utility bill, since winter shutoffs are commonly prohibited, and budget counseling and extended payment terms are often provided. However, the provision of this credit can result in significant arrearages for utilities. In Pennsylvania alone, utilities carry an ongoing arrearage in excess of $90 million. Nationwide, utilities often have bad debt write-offs in excess of 0.5% of net revenues (McBride et al. 1988; NCLC 1988b).

History of Low-Income Weatherization Services

The initial low-income energy assistance programs were created in response to the rapid escalation in energy prices and supply constraints brought about in part by the Arab oil embargo of 1973–1974. The first agency to address the problem was the Community Services Administration (CSA), successor to the Office of Economic

Opportunity. In 1974, CSA authorized its Community Action Agency (CAA) grantees to spend up to 10% of their general operating funds on fuel assistance and energy conservation, then known as Crisis Intervention (Sweet and Hexter 1987). The fuel assistance program eventually became the Low-Income Home Energy Assistance Program (LIHEAP), administered by the U.S. Department of Health and Human Services.

The residential energy conservation program under CSA became known as weatherization. A low-income weatherization program, the Weatherization Assistance Program, was established within the Federal Energy Administration (now the U.S. Department of Energy) under the Energy Conservation and Production Act of 1976. For several years both CSA and the U.S. Department of Energy (DOE) funded weatherization programs. With the elimination of CSA in 1981, weatherization was funded entirely by DOE.

The weatherization program evolved significantly over time. In the initial DOE program, at least 90% of grant funds were to be spent for weatherization materials. No more than 10% could be spent on administration. All labor was to be provided by trainees under the Comprehensive Employment and Training Act (CETA). The program gave priority to weatherizing the residences of the elderly and disabled. Rental units could be weatherized if the benefits of weatherization accrued primarily to the tenant rather than the landlord, and specifically, rents could not be raised on the basis of improvements resulting from weatherization.

The weatherization program regulations were amended in 1979 to raise the materials expenditure limit from $400 to $800 per dwelling unit, permit furnace repairs and modifications, and increase the income limit for program eligibility from 100% to 125% of the poverty level. Further amendments were promulgated from time to time until by 1985 the states, at their discretion, could raise the program income eligibility limit to 150% of the poverty level and spend an average of $1,600 per dwelling unit for weatherization with at least 40% of this amount spent for materials. By 1985, the CETA program had ended, and CAAs either employed their own crews or hired contractors to provide weatherization labor. Furthermore, in 1985, furnace and boiler replacement was permitted, a $150 per dwelling limit on incidental repairs was removed, and agencies were permitted to reweatherize dwellings completed in the early years of the program. Specific regulations for weatherizing multifamily rental units also were implemented.

Funding for low-income weatherization has come from a variety of federal sources in addition to DOE in the past few years.

One major source of weatherization funding is LIHEAP. Funding for LIHEAP, although dramatically reduced since 1985, still exceeds $1 billion annually. States can, and typically do, transfer up to 15% of the LIHEAP block grant funds to weatherization. Oil overcharge restitution funds have also been a major source of weatherization funding in recent years. Since 1986 roughly $3.2 billion has been distributed to the states for energy-related activities as the result of legal settlements stemming from alleged violations of price controls by oil companies at the time of the Arab oil embargo. Over $627 million of these funds has gone for low-income weatherization, with an additional $623 million allocated to LIHEAP as of August 1989 (NCLC 1989). Figure 6-1 shows federal weatherization funding levels in nonadjusted dollars from the start of the program in 1974 to the present.

In recent years some utilities—on their own or at the direction of state regulatory commissions—have initiated low-income weatherization programs. These programs have typically operated independently from the DOE-funded programs and followed their own

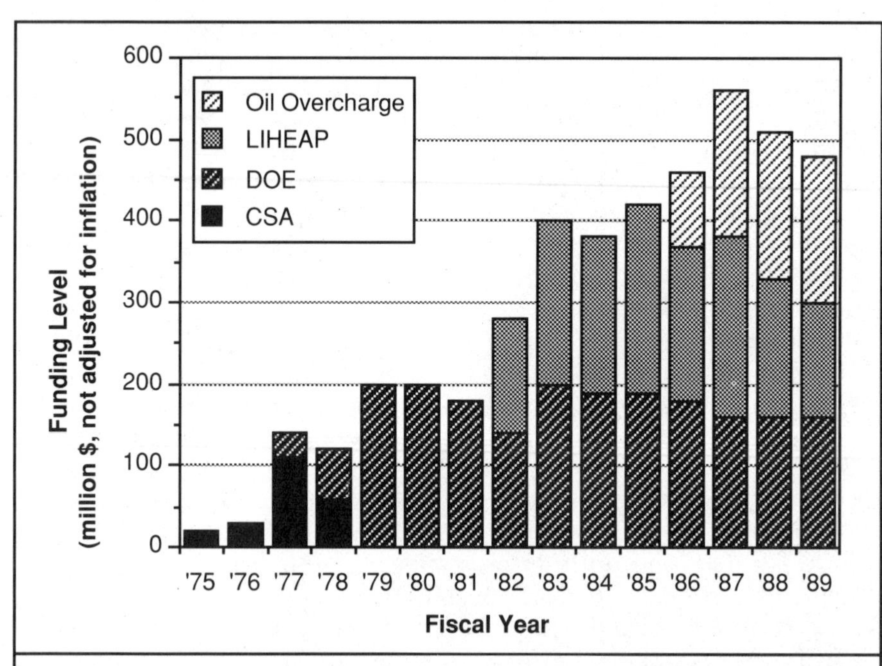

Figure 6-1. Combined weatherization program annual funding levels, FY 1975–1989.

operating procedures and guidelines. California utilities established low-income weatherization programs in the early 1980s at the direction of the California Public Utilities Commission. Similarly, Wisconsin, Pennsylvania, and Connecticut all have commission-directed, utility-funded, low-income weatherization programs. The Bonneville Power Administration, which supplies electricity to utilities in the Pacific Northwest, has operated a multimillion dollar low-income weatherization program since 1982. Nationwide, total annual budget levels exceed $50 million (about 10% of the federal program level) for utility-funded weatherization programs, and are growing rapidly.

Funding for low-income weatherization is now at a critical juncture. Utility funding, though growing rapidly, is still relatively small compared to federal funding, which is slowly declining (see Figure 6-1). In addition, the bulk of the anticipated oil overcharge funds has already been distributed. A precipitous drop in federal low-income weatherization funding could be looming at a time when the need is as great or greater than ever.

The need for low-income weatherization has always outpaced the resources available for the program. There are currently over 20 million households in the United States living below 150% of the poverty level and theoretically eligible for weatherization (NCLC 1989). Only 2 million households had been weatherized by 1989 under the DOE program. A substantial number have also been weatherized under other programs (utilities, oil overcharge, and so on) in recent years. At current DOE weatherization production levels of 300,000 units per year, it would still take 50–60 years to weatherize the eligible households. At an average cost of $2,000 per home, weatherizing all eligible households would result in a $30+ billion expenditure in current dollars. At an average annual energy savings of 20 million Btu per year per unit weatherized, the resultant energy savings would be about .3 quads (quadrillion Btu) per year or about 0.4% of total current U.S. energy consumption.

DOE Weatherization Assistance Program

Community Action Agencies have been and continue to be the cornerstone of the DOE Weatherization Assistance Program. To ascertain the current status of low-income weatherization, NCAT mailed a questionnaire to each of the 991 Community Action Agencies and sought responses from a sample of 288 of these agencies randomly selected to ensure representative data. Any sampled CAA that did not return the questionnaire received a follow-up phone call. From the sample, 168 CAAs (17% of the population) responded (NCAT 1990).

Community Action Agencies are diverse but relatively small agencies. They were originally created by the Office of Economic Opportunity in the 1960s as frontline antipoverty agencies. They have survived assaults on their program funding by both the Nixon and Reagan Administrations and even survived the elimination of the Community Services Administration in 1981. Of the CAAs in the NCAT sample, 8% had annual agency budgets of less than $250,000, 23% had budgets between $250,000 and $1 million, 50% had budgets between $1 million and $4 million, and 19% had budgets exceeding $4 million. CAAs appear to be stronger in rural areas. Of the CAAs, 20% served urban clients, 43% primarily served rural clients, and 38% served both.

Weatherization has always been a major program for CAAs. Approximately 85% of the sampled CAAs operated a weatherization program. DOE is the principal source of funding for these programs, with 93% of the CAAs receiving funding from this source. Local governments provide funding for 10% of the CAAs. Fuel funds, a source of private sector and utility company funds, support weatherization programs in 36% of the CAAs in the sample. Utilities fund weatherization programs in 23% of the CAAs in the sample, while 21% of the CAAs depend on other sources of weatherization funding. Utilities appear to be a partially untapped source of weatherization funds for CAAs. One-half of the CAAs in the sample reported that utilities in their service areas operated weatherization programs, but less than one-third of the CAAs received utility funding for their weatherization programs.

The number of units weatherized annually by CAAs varies widely. The mean number of units weatherized annually by the agencies in the sample was 300, with the number per agency ranging from 12 to 3,940. Based on the NCAT sample, CAAs together probably weatherize about 300,000 units per year. (The DOE estimate for program year 1988 was 265,000 units [Sheladia Associates 1988].)

The majority (91%) of the CAAs that operate weatherization programs address single-family detached homes, 80% address mobile homes, 57% address small multifamily buildings, 23% address row houses, and 17% address large multifamily apartment buildings. The majority of all housing units weatherized by the CAAs are single-family, detached homes (66%). Mobile homes represent 19%, small multifamily units (2–4 units/building) 10%, row houses 3%, and units in large multifamily buildings 2%. Nearly all (97%) CAAs in the sample weatherized rental units.

In the early days of low-income weatherization, all of the labor was provided under the CETA program. As the CETA program was phased out, CAAs developed other means of providing weatherization labor. Today, 39% of the CAAs in the sample hire their own weatherization crews, 44% hire contractors, and 17% have both crews and contractors. Labor allocation is an important issue for CAAs. Under DOE regulations, no more than 60% of weatherization funds may be spent on labor costs. This means that some agencies, particularly those that rely on weatherization contractors, may emphasize material rather than labor-intensive weatherization techniques, regardless of cost-effectiveness.

Access to training and the associated technology transfer can greatly influence the state of the practice of low-income weatherization. Essentially, all of the respondents in the survey received some type of weatherization training, most commonly provided by the state weatherization agency. Training typically covers both the techniques of weatherization and the operation of the program itself. Some utilities also provide technical weatherization training. A few states have established weatherization training centers. Pennsylvania, for example, has established a weatherization training center at the Williamsport College of Technology, a branch of Pennsylvania State University. Only a few of the CAAs sampled indicated that they received up-to-date information on the latest advancements in weatherization technology.

Funding limits on weatherization and incidental repairs also influence the current state of the practice of low-income weatherization, as do cost-effectiveness criteria. Program priorities and funding limits are often at cross purposes. Current DOE regulations require that expenditures average less than $1,600 per weatherized home, but there is no limit on how much can be spent per home so long as the average stays within limits. Given the enormous backlog of homes to be weatherized, many states and CAAs give preference to customers with high energy bills. These high users typically require extensive weatherization costing well in excess of $1,600 per home. The high users typically also save the most energy after weatherization and thus receive the most financial benefit from weatherization. Thus, by addressing the neediest customers first, CAAs' average expenditure per home can easily exceed the $1,600 limit.

Just over one-half of the survey respondents indicated that they incorporate a cost-effectiveness analysis into their weatherization program. Cost-effectiveness criteria help direct weatherization programs to install effective measures and serve as a guide when financial resources are limited. Cost-effectiveness criteria can be applied to the individual measure, individual home, or entire program. These criteria generally are more advanced than the simple prescriptive approaches of some priority lists. Cost-effectiveness criteria may conflict with the 60/40 labor-to-materials rule, however, since they do not distinguish between labor-intensive or materials-intensive weatherization approaches.

The lack of integrated funding—specifically, different regulations by different funding sources—is an issue affecting the practice of low-income weatherization. Most of the survey respondents received weatherization funding from multiple sources. While oil overcharge funds and LIHEAP transfers used for weatherization typically fall under DOE weatherization regulations, funds from private sources such as fuel funds and utilities do not. As funding from these private sources grows, so does the potential for conflicts in program approaches, policies, and regulations. In the worst case, CAAs have reported making multiple visits to the same house to conduct duplicate audits because weatherization funding came from multiple sources. One response to this problem is to package energy services into standard programs that deliver the same comprehensive services irrespective of funding source. The state of New York and the cities of Philadelphia and Minneapolis are among the leaders in energy services packaging.

Utility Low-Income Weatherization Programs

In contrast to the federal government, public utility commissions and utilities are only now beginning to create comprehensive conservation programs for low-income ratepayers. However, three comprehensive utility-funded low-income weatherization programs did begin in the early 1980s. California utilities began targeting weatherization services to low-income households in 1980. In 1982, the Public Service Commission of Wisconsin directed Class A utilities in the state to design and implement a program to weatherize the homes of low-income ratepayers. Also in 1982, the Bonneville Power Administration began a low-income weatherization program for utilities in Washington, Oregon, Idaho, and Montana.

The California weatherization program originated in 1980 when Pacific Gas and Electric Company offered a zero-interest weatherization loan program for all residential ratepayers. After a study found little low-income participation in the zero-interest loan program, the California Public Utilities Commission ordered the state's utilities to offer free, direct weatherization services to low-income households. California utility programs contracted with community-based organizations, local governments, and insulation contractors to install attic insulation, weather stripping, caulking, water heater blankets, low-flow shower heads, and duct wrap in dwellings occupied by low-income residents. Through 1988, California utilities have weatherized almost 250,000 such dwellings at a cost of $140 million (Barnett 1988).

In 1982, the Public Service Commission of Wisconsin initiated an investigation to determine if a utility-financed, low-income weatherization program was warranted. The investigation found that the percentage of income spent by low-income natural gas customers on space heating was three times more than the percentage spent by higher-income ratepayers, that low-income customers had little or no income to invest in conservation measures, and that providing low-income customers with weatherization would give all ratepayers an opportunity to install conservation measures (PSC Wisconsin 1982).

As a result, the commission required the major natural gas and electric utilities to provide direct weatherization to low-income customers at no charge. Since 1983, Wisconsin utilities have weatherized 40,000 low-income households at a cost of $70 million. The utilities deliver their own weatherization services and coordinate jobs with the state weatherization providers. Measures installed include high-efficiency furnaces, attic insulation, wall insulation, compact fluorescent light bulbs, high-efficiency refrigerators, and some low-cost weatherization items. A recent evaluation found natural gas savings ranged from 15% to 20% (Horowitz and Degens 1987).

In 1980, Congress passed the Northwest Power Act, which created the Northwest Power Planning Council and made the Bonneville Power Administration (BPA) responsible for meeting the power needs of Northwest utilities requesting BPA service. The act gave conservation priority as a source of new energy supply, and since 1980, conservation programs have added nearly 300 megawatts (MW) to BPA's energy resources (BPA 1990). One component of BPA's conservation program is low-income weatherization. The program began in 1982 and permits BPA to pay up to 100% of the actual cost for local utilities to weatherize homes (BPA 1986). Up to $150 for home repairs was also permitted. The utilities often use CAAs as weatherization providers BPA weatherization is typically quite extensive, with insulation, storm windows, and infiltration reduction measures commonly installed. A recent evaluation of the program found the average cost of weatherization to be $1,847 per residence with an average savings of 8.9% (Haeri, Bronfman, and Lerman 1988).

A few other state regulatory commissions have enacted or are pursuing mandates for utility involvement in low-income weatherization. The Public Utilities Commission of Ohio addressed the issue in an investigation of long-term solutions to the disconnection of gas and electric service in the winter. Low-income weatherization was thought to be a way to reduce arrearages and potentially address the low-income bill payment problem. However, the commission did not order a broad low-income weatherization program because it determined that it lacked sufficient statutory authority. The commission also believed that the costs and benefits of such a program would be inequitable for ratepayers, that there was a potential for industrial load shift as a result of rate increases associated with the program, and that the utilities did not have the expertise to deliver a low-income weatherization program. The commission did accept stipulations signed by the major utilities of Ohio, which provided an energy audit and weatherization kit to electric low-income households and an experimental weatherization demonstration program for low-income gas customers. The stipulations began in 1985 and expired in 1990 (PSC Ohio 1985).

In 1987, the Pennsylvania Public Utilities Commission adopted regulations establishing a comprehensive low-income weatherization program. The state's 14 gas and electric utilities were ordered to provide these services at no cost to low-income customers. The commission took this action after determining that low-income customers had been required to pay via rates for many utility conservation incentive programs without having the financial ability to participate in those programs. The intent of the low-income program is to reduce the residential demand for natural gas and electricity and to reduce the uncollectible accounts and termination expenses of the utilities by enabling low-income customers to reduce their energy consumption (Penn. PUC 1987).

A 1987 ruling by the Connecticut Public Utilities Commission provided the foundation for utility involvement in low-income weatherization. When Northeast Utilities' Connecticut Power and Light Company requested to reduce its load management and conservation activities, the commission responded by requiring greatly enhanced programs including many focused on low-income customers. The commission determined that low-income customers suffer when a utility does not develop programs to mitigate their energy burden. The ruling reemphasized the need for least-cost planning and stated that conservation and load management were critical in meeting current and future capacity needs. The company was required to work collaboratively with several intervenors to establish an enhanced weatherization program for low-income customers. To reflect the alliance of utilities, CAAs, and public agencies, the program was named Weatherization Residential Assistance Partnership (WRAP). Conn Save, a nonprofit residential auditing service, administers the program. The CAAs provide a comprehensive array of 31 conservation measures (gas, electric, and oil). The program budget for 1989 was $3.64 million, with $2.56 million in public funds and $1.08 million in utility funds. The 1989 weatherization goal was 4,311 units (Ellison et al. 1989).

A proceeding to determine whether major natural gas and gas-and-electric utilities should establish a low-income energy-efficiency program is presently before the State of New York Public Service Commission. The commission staff has argued that a program targeting high-usage, payment-troubled customers could significantly benefit both utilities and ratepayers and enhance the commission's effort to promote energy conservation and least-cost planning. Program benefits beyond conservation gains are expected. These benefits include a reduction in arrearages by lowering the total energy demands of households and increased ability of low-income customers to pay their utility bills. Ratepayers and utilities would also benefit from reduced costs of bill collection, service termination, service reconnection, and complaint handling. The proceeding should reach closure in 1990 with an expected ruling on the need for utility low-income weatherization in the state of New York (N.Y. PSC 1989).

Technical State of the Art

The Focus on Program Improvement

During the past 12 years, the performance of weatherization programs has improved dramatically. In the late 1970s, the programs were mostly no cost/low cost and installed inexpensive measures that provided little energy savings. In the early 1980s, attic insulation, storm windows, and caulking and weather stripping were installed most frequently. While many advocates and administrators of low-income weatherization programs believed the programs were saving approximately 20–25% of energy use, some well-designed evaluations indicated savings as low as 6–8% (Goldberg, Jaworski, and Tallis 1984; Carmody 1986).

Both the technical and the program potential for energy savings, however, are much higher than either the results measured in past programs or the results estimated by weatherization advocates. Engineering calculations, energy audits, and carefully designed pilot/demonstration projects all indicate that the weatherization of low-income homes should provide substantial energy savings. For example, in a demonstration project, the National Bureau of Standards (NBS) retrofitted building shells and heating systems in more than 200 low-income houses at 12 sites across the United States. Homes that received both types of retrofits consumed an average of 41% less space heating fuel at an average cost of $1,862 per house (Crenshaw and Clark 1982).

While the results of the early evaluations were disappointing, the results of the NBS demonstration project indicated that there was a large potential for energy savings. These facts caused managers, advocates, and technical consultants to seriously question the effectiveness of existing programs. What followed was a period of experimentation where the standard of performance was enhanced through pilot programs and field tests. During the last five years, there has been a dramatic increase in the level of technical sophistication in low-income weatherization programs. This has been accomplished through (1) the installation of a wider range of more cost-effective energy conservation measures, (2) the use of more effective energy audits and measure selection protocols, (3) the increased use of diagnostic tools, (4) education programs for clients and customers, (5) better training for program personnel, and (6) enhanced feedback from program evaluations. Recent efforts have focused on having the state of the practice catch up to the state of the art.

Weatherization Measures

The current emphasis in weatherization programs is on building science and integrated approaches to buildings and their occupants. A broader mix of energy conservation measures addressing the building shell, heating and air-conditioning systems, water heaters, infiltration, lighting, occupant behavior, and health and safety concerns are being implemented. Table 6-1 summarizes the major measures currently installed by CAAs in the DOE Weatherization Program (NCAT 1990). Heating systems in particular are being addressed more frequently. In 1979 only 33% of programs installed any type of furnace retrofit (NCAT 1979), while in 1990, 56% of programs had heating system testing and repair services, and 48% retrofitted or replaced heating systems. The number of programs incorporating other more cost-effective weatherization measures, such as wall insulation, lighting, and water heating retrofits, also

Table 6-1. Percent of CAA programs installing weatherization measures.

Measure	Percent
Caulking/Weather Stripping	96%
Hot Water Pipe Insulation	96%
Ceiling Insulation	94%
General Infiltration Reduction	94%
Replacement Windows and/or Doors	94%
Water Heater Insulation	94%
Floor/Basement Insulation	91%
Storm Windows	68%
Wall Insulation	59%
Furnace Testing and Repair	56%
Heating System Retrofit/Replacement	48%
Hot Water Flow Restrictors	46%
Zone Heating/Warm Room	23%
Storm Doors	23%
Lighting Retrofits	15%
Mechanical Cooling Repair/Replacement	6%

Source: NCAT 1990.

has increased. Improved installation techniques, such as dense pack cellulose wall insulation, are incorporated more frequently as well.

Unfortunately, many programs are still installing measures that adversely affect program cost-effectiveness. Most notable among these measures are replacement windows or doors or both (94% of programs) and storm doors (23% of programs). While these measures save some energy, they have not been shown to be cost-effective in low-income weatherization programs (Hewitt et al. 1984). Other measures, such as adding storm windows, may be cost-effective in electrically heated homes but not in homes with gas heat (Hewitt et al. 1984; Gregory 1987; Gregory 1989a).

The measures installed by utility programs vary significantly, from low-cost measures, such as water heater insulation and weather stripping, to attic and wall insulation. In the more comprehensive programs, utilities have generally relied on mechanical system retrofits and replacements. In Wisconsin, utilities replace the heating system in approximately 35% of buildings (Horowitz and Degens 1987). The popularity of this measure with Wisconsin utilities is due to the significant energy savings associated with replacements, the improved safety and reliability of new heating systems, the long estimated life of the measure, and the ease of service delivery using heating contractors. Many utilities also have comprehensive operation and maintenance programs for mechanical systems. While maintenance may not save as much energy as replacement, it is generally more cost-effective (Horowitz and Degens 1987).

Measure Selection Protocols and Energy Audits

Many programs have improved their energy audits and measure selection protocols. As auditors assessed a greater number of building components and considered a wider range of conservation measures, it was necessary to develop and use more sophisticated protocols. Many of the newer protocols allow auditors to examine unique aspects of individual buildings and recommend site-specific measures. The traditional priority lists have been improved by incorporating more options and using decision rules in many instances, particularly for HVAC system measures (Hewitt et al. 1984; Kushler and Witte 1988; Shen et al. 1990). Many CAAs are using calculation type audits (23% of programs) and computerized audits (15%) (NCAT 1990). The combination of a wider range of possible measures and more sophisticated measure selection protocols has led to an increase in regional and site-specific methodologies.

Diagnostic Tools

Diagnostic tools are being used more frequently in both DOE and utility programs. Tools, such as blower doors, enable trained and experienced technicians to make better decisions about how a building performs and which energy conservation measures would be most effective. Ten years ago, few CAAs had heard of blower doors. Today, blower doors are used in 42% of CAA programs, and their functions in these programs have evolved. Originally, blower doors were just used as training tools or for quality control. Now they are used in many programs by field personnel on every job to measure the initial leakage rate of the building, help locate leakage areas, and provide feedback on the impact and cost-effectiveness of the air-sealing work. In addition, measurements made with the blower door help determine when a building is too tight or tight enough (Schlegel 1990). Other diagnostic tools are also being used more frequently. For example, heating system test equipment is used by 43% of programs, and infrared cameras are used by 13%.

The Impact of Technical Improvements on Program Personnel

The increase in technical sophistication has had a strong impact on program personnel. Field assessment staff have evolved from estimators to energy auditors to energy and building science technicians. Five years ago, many auditors were mainly estimators, determining how many bags of insulation to put on the truck or what size a storm window should be. Today, many auditors are trained and experienced technicians, assessing complex interrelationships of energy use and building dynamics and determining the most effective measures for a given building. Many auditors use diagnostic tools, engineering models, and energy audits to assist them in the process.

As the program has addressed more weatherization options, field personnel have enhanced their technical and communication skills. Depth and breadth have both increased dramatically—the job is no longer single purpose. Once auditors are trained and experienced, they become very valuable to the program. Other program personnel, from installers to program managers, have also increased their skill levels and their technical understanding of energy and buildings. Some programs have implemented certification programs to ensure that only trained and experienced people are delivering services, to recognize the enhanced status of the work, and to formalize the career ladder process.

Program Evaluation

Program evaluation has played a key role in improving program performance. Evaluations have provided the necessary feedback to program managers and personnel on what worked, what did not, and where improvements could be made. Several state programs, especially in Michigan, Ohio, and Minnesota, have used evaluations to chart a course for improved program performance and check on progress along the way. BPA has also made a commitment to ongoing evaluation of its low-income weatherization program (Haeri, Bronfman, and Lerman 1988).

With the recent history of reduced federal funding and the current atmosphere of questioned program performance, low-income weatherization programs must be accountable. Evaluation can quantify performance and indicate where further improvements need to be made. Evaluation and program monitoring will likely become even more important and widely practiced in the near future. Of the CAAs responding to the NCAT survey, 86% indicated that they have evaluated the performance of their program. Since these are self-reported results, questions remain about the evaluation methodologies employed and their rigor. In addition, many of these evaluations have not been published and therefore have not been reviewed by peers. Other studies have indicated that many states have yet to evaluate their DOE-funded program, particularly in cooling climates, and have questioned the rigor and accuracy of some of the methodologies used for evaluations to date (Meridian Corporation 1988).

Current Program Performance

Due to increased technical sophistication, the performance of weatherization programs has improved over the last eight years, but the impact of these improvements on energy savings and cost-effectiveness has not yet been evaluated on a national scale. Individual state and utility programs have been evaluated, however; 13 such evaluations are summarized below and serve as reviews of the past and current performance of low-income programs.

Performance in Gas-Heated Homes

Table 6-2 (pages 178 and 179) summarizes the results of 12 recent evaluations of weatherization programs addressing natural-gas–heated homes in heating climates (Banerjee and Goldberg 1985; Cambridge Systematics 1988; Carmody 1986; Goldberg, Jaworski, and Tallis 1984; Gregory 1987 and 1989b; Horowitz and Degens 1987; Kushler, Witte, and Stanley 1987; Kushler and Witte 1988; McCold et al. 1988; Shen et al. 1986; Shen et al. 1990).[1] The studies include both utility and DOE programs. Most of these studies either measured the effect of improved program delivery systems and more cost-effective measures, or discussed how the impact or cost-effectiveness of the program might be increased (Schlegel and Pigg 1990).

Annual energy savings ranged from 6% to 23% (58–791 therms) (Figure 6-2). Early DOE programs (WISC LI and MINN) saved 6–10% (79–118 therms). These programs installed a few measures selected from a short priority list. Programs that addressed a wider range of opportunities and installed more comprehensive packages of measures achieved greater savings. Savings for eight of these comprehensive programs in the Midwest ranged from 15% to 23% (207–348 therms). Energy savings for these eight programs remained in this range despite a variety of approaches to weatherization. The program that achieved the highest energy savings (791 therms) was a home repair and weatherization program targeted to extremely high energy users (MICH HRW).

Program spending ranged from $1,450 to $3,461 per building, with most programs between $1,500 and $2,250. For DOE-funded

---

1. See Table 2 for the abbreviations used to refer to these studies in the text and in Figures 2 and 3.

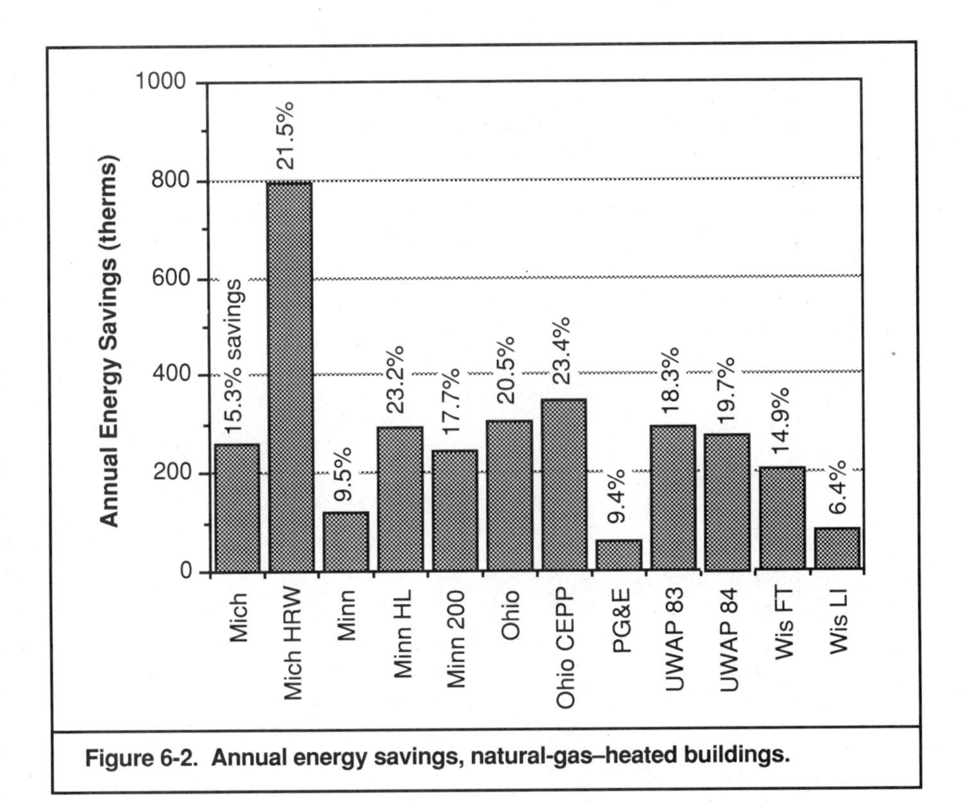

Figure 6-2. Annual energy savings, natural-gas–heated buildings.

programs, this similarity in cost likely reflects the DOE program guidance of $1,600 average spending per building.

Cost-effectiveness, as represented by the cost of conserved energy (CCE), ranged from $4.22 to $27.44/MMBtu (Figure 6-3, page 180). The nine programs that installed comprehensive packages of measures were more cost-effective, with a range of $4.22 to $10.37/MMBtu, with eight programs between $4.22 and $7.46/MMBtu. These figures compare to CCEs of about $12 to $27/MMBtu for the earlier programs, which used short priority lists, and to retail rates for natural gas of about $4.50 to $6.00/MMBtu. The cost-effectiveness of these programs will continue to improve as programs incorporate technical improvements, target services where there are greater opportunities, achieve higher savings, and become more efficient in their program delivery systems (Schlegel and Pigg 1990).

Program Potential

The level of measured energy savings (change in energy intensity)

Table 6-2. Results of evaluations of low-income weatherization programs, natural-gas–heated buildings.

	MICH Kushler 1988	MICH HRW Kushler 1987	MINN Carmody 1986	MINN HL Shen 1986	M200 Shen 1990	OHIO Gregory 1987	OH/CEPP Gregory 1989b	PG&E Cambridge 1988	UWAP 83 Banerjee 1985	UWAP 84 Horowitz 1987	WIS FT McCold 1988	WIS LI[a] Goldberg 1984
Year of Weatherization	1986	1985	1984	1985	1988	1985	1987	1986	1982/3	1984/5	1986	1982
Sample Size	173	158	221	14	208	1,083	316	9,481	1,357	483	20	353
Building Size (ft^2)	1,250[b]	1,475	912	988	1,346	1,222	1,222[c]	1,200[d]	1,274[e]	1,274[f]	1,286	1,042
HDD_{60}[g]	5,475[h]	5,475[h]	6,824[i]	6,824[i]	6,824[i]	4,491[j]	4,491[j]	1,643[k]	6,416[l]	6,416[l]	6,352[m]	6,416[l]
Program Cost ($)	1,562[n]	3,461	1,450	3,144	1,571	1,980[n]	2,049[o]	NA	2,134	1,894[p]	1,603[q]	2,250
Pre-NAC[r] (therms)	1,705[s]	3,681	1,245	1,256	1,375	1,487	1,487[t]	620	1,611	1,392	1,393[u]	1,242
Post-NAC (therms)	1,444	2,890	1,127	964	1,132	1,182	1,139	562	1,316	1,118	1,186[v]	1,163
Savings (therms)	261	791	118	292	243	305	348	58	295	274	207	79
Savings (%)	15.3	21.5	9.5	23.2	17.7	20.5	23.4	9.4	18.3	19.7	14.9	6.4
CCE[w] ($/MMBtu)	5.77	4.22	11.81	10.37	6.23	6.64	6.01	NA	6.97	6.66	7.46	27.44
Energy Intensity ($Btu/ft^2\text{-}HDD_{60}$)												
PRE	24.9	45.6	20.0	18.6	15.0	27.1	27.1	31.3	19.7	17.0	17.1	18.6
POST	21.1	35.8	18.1	14.3	12.3	21.5	20.8	28.4	13.7	14.7	14.5	17.4
CHANGE	3.8	9.8	1.9	4.3	2.6	5.6	6.3	2.9	3.4	2.3	2.5	1.2

(continued on next page)

Table 6-2 *(continued).*

a. Program abbreviations stand for the following:

MICH: State of Michigan
MICH HRW: State of Michigan Home Repair and Weatherization
MINN: State of Minnesota
MINN HL: Minnesota High-Level Weatherization Project
M200: Minnesota 200 Demonstration Project
OHIO: State of Ohio
OH/CEPP: Ohio Client Education Pilot Program
PG&E: Pacific Gas and Electric, Gas Buildings
UWAP: Utility Weatherization Assistance Program (Wisconsin Utilities)
WIS FT: Wisconsin Field Test
WIS LI: Wisconsin Low-Income Weatherization Assistance Program

b. Estimated
c. Estimated from previous Ohio study
d. Estimated
e. Estimated from UWAP84 data
f. Estimated based on building volume data
g. Heating degree-days at a base temperature of 60°F
h. Population-weighted Heating Degree Days for Michigan
i. For Minneapolis, Minnesota
j. For Columbus, Ohio
k. For San Francisco, California
l. Population-weighted Heating Degree Days for Wisconsin
m. For Madison, Wisconsin
n. Not reported in evaluation; obtained by direct communication with author
o. Based on reported $69 cost for education component, plus $1,980 reported for previous Ohio evaluation
p. Based on reported $1,594 direct costs, plus an estimated $300 in program costs
q. Based on reported $1,303 direct costs, plus an estimated $300 in program costs
r. NAC: normalized annual consumption
s. Estimated from reported heating energy consumption of 1,345 therms (360 therm baseload)
t. Estimated from previous Ohio study
u. Estimated based on reported 1,033 therms heating energy consumption (360 therm baseload)
v. Estimated based on reported 826 therms heating energy consumption (360 therm baseload)
w. Cost of conserved energy calculated using a real discount rate of 5% and a measure lifetime of 15 years

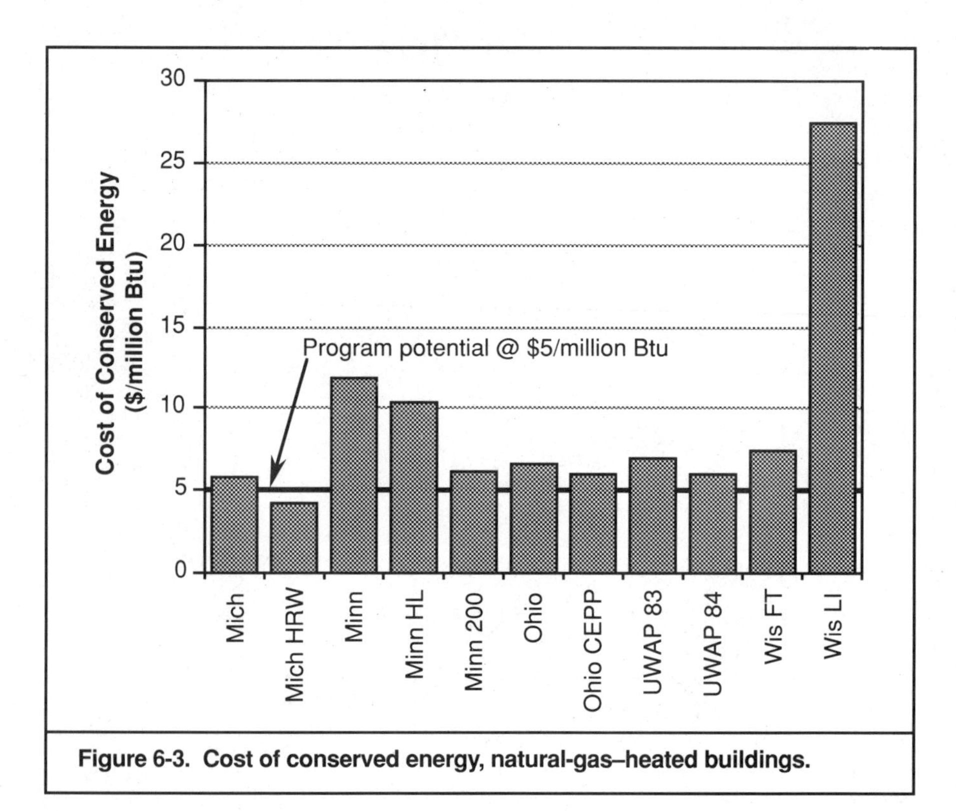

Figure 6-3. Cost of conserved energy, natural-gas–heated buildings.

is correlated with the potential for energy savings, defined as pre-retrofit energy intensity (Figure 6-4). Programs that operate in energy intensive housing stock could achieve further reductions in energy consumption, while those operating in less energy intensive housing stock may be approaching their near-term potential; near-term potential energy savings for the 11 weatherization programs were estimated to range from 300 to 1,400 therms (Schlegel and Pigg 1990). Pilot projects for two of the weatherization programs demonstrated that increased energy savings could be achieved either either at a cost comparable to that of programs lacking the new techniques or at an attractive marginal cost-effectiveness. In addition, programs should be able to achieve a cost of conserved energy of \$5.00/MMBtu (Schlegel and Pigg 1990). This program potential for cost-effectiveness is shown in Figure 6-3.

Performance in Electric Homes

Data on the performance of low-income weatherization in electric homes is more limited than that on homes heated by natural gas.

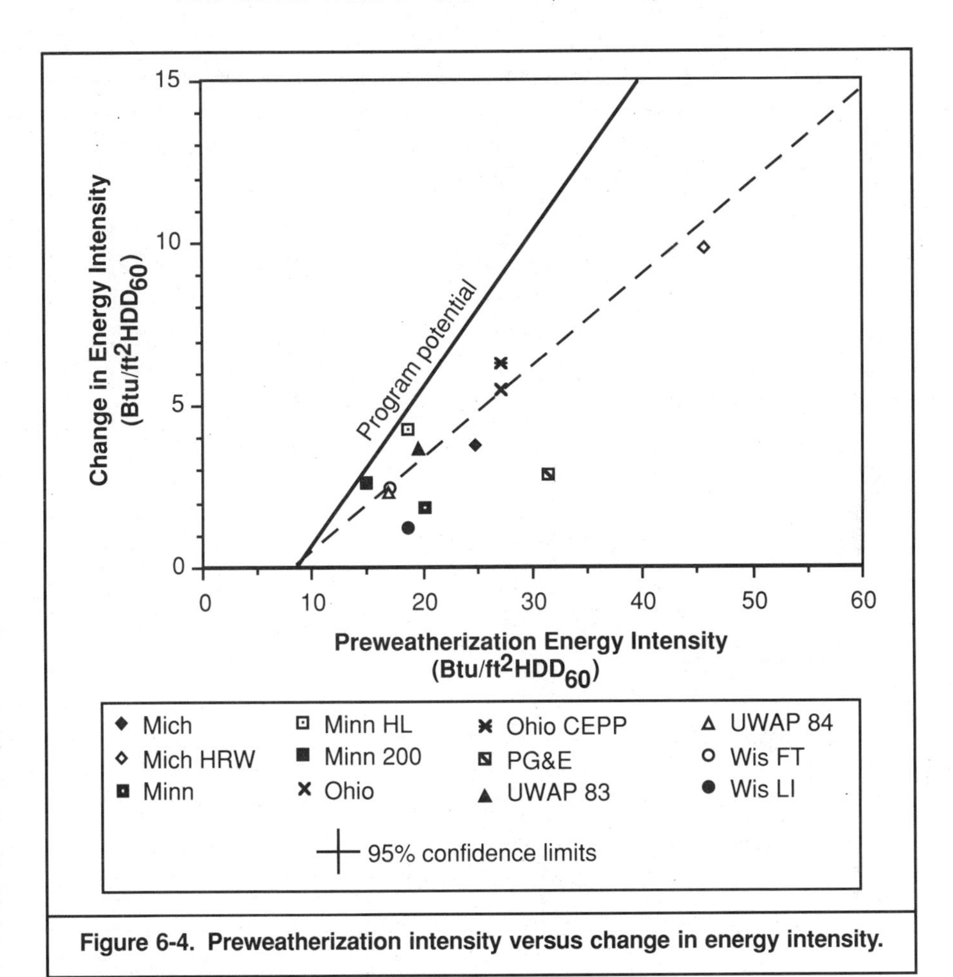

Figure 6-4. Preweatherization intensity versus change in energy intensity.

Table 6-3 (next page) shows results for two programs in the western United States (Haeri, Bronfman, and Lerman 1988; Cambridge Systematics 1988). Annual energy savings ranged from 36 kWh to 2,475 kWh, or from 0.76% to 8.9% (Cambridge Systematics 1988; Haeri, Bronfman, and Lerman 1988). The lower figure for energy savings is for homes in California that did not have electric heat or air-conditioning; the higher figure is for electrically heated homes in the Pacific Northwest.

Future Directions

Low-income weatherization programs are facing many issues and

Table 6-3. Results of evaluations of low-income programs, electric buildings.

	BPA	PG&E HEAT	PG&E HT/AC	PG&E AC	PG&E BASE
		Elec. Heat, No AC	Elec. Heat, Plus AC	Non-elec. Heat, Plus AC	Non-elec. Heat, No AC
	Haeri 1988	Cambridge 1988	Cambridge 1988	Cambridge 1988	Cambridge 1988
Year of Weatherization	86	86	86	86	86
Sample size	161	260	147	1,442	7,893
HDD_{60}[a]	4,920[b]	1,643[c]	1,643	1,643	1,643
Program cost ($)	1,847	NA	NA	NA	NA
Pre-NAC[d] (kWh)	22,141	9,129	10,267	6,139	4,810
Post-NAC (kWh)	19,666	8,770	9,697	5,829	4,774
Savings (kWh)	2,475	359	571	311	36
Savings (%)	8.9	3.93	5.56	5.06	0.76
Cost of conserved energy[e] ($/MMBtu)	21.07[f]	NA	NA	NA	NA

a. Heating degree-days at a base temperature of 60°F
b. Sample weighted degree days adjusted to base 60°F
c. For San Francisco
d. NAC: normalized annual consumption
e. Cost of conserved energy calculated using a real discount rate of 5% and a measure lifetime of 15 years.
f. Haeri et al. (1988) reported a CCE of 42.2 mills/kWh assuming a 31-year financing term, 8.35% BPA financing rate, and 5% annual inflation. This is equivalent to $12.31/MMBtu.

challenges in the near future. The key policy, program, and technical issues are discussed below. These issues include: (1) the need for increased funding, (2) future utility involvement in low-income weatherization, (3) integration of programs and services, (4) targeting of services, (5) low-income rental issues, (6) technical sophistication, (7) energy education, (8) impact on program delivery systems, and (9) program evaluation and monitoring.

The Need for Increased Funding

When the NCAT survey respondents were asked to identify the single most important thing that could be done to improve the effectiveness of low-income weatherization, the overwhelming response was to provide more funding. Low-income programs are faced with a situation in which the need for weatherization far surpasses the ability to provide the service. To date, only 2 million of the more than 20 million eligible households in the United States have been weatherized (NCLC 1989). At an average cost of $2,000 per home, the cost to weatherize the remaining18 million households would be $36 billion in current dollars.

Increased funding for weatherization can thus be argued based on both need and societal benefit. Establishing a national goal of weatherizing the low-income housing stock in the next ten years would require a six-fold increase in annual production. Such a dramatic increase may not be practical or possible immediately, given the existing weatherization support infrastructure. Yet, if weatherization funding from all sources was increased 40% per year for the next five years and then held constant, the job would be done in 13 years. Weatherizing the nation's low-income housing stock is not an impossible task but simply requires a national mandate.

Future Utility Involvement in Low-Income Weatherization

More regulatory commissions are expected to consider the issue of encouraging or mandating utility low-income weatherization programs financed with ratepayers' dollars. There are three major reasons for this expectation.

First, as the need for conservation and least-cost planning is more widely recognized, commission proceedings on a full range of conservation and load management activities will produce a mandate for increased expenditures and programs by utilities, including programs targeted to the low-income sector. Utility low-income programs will be evaluated by both the energy savings to the household and the benefits to the utility system. For many utilities, it will be difficult for low-income weatherization programs to meet cost-effectiveness tests based solely on utility avoided costs. It will also be necessary to quantify other aspects of the program, such as arrearage reduction, bill payment behavior, and savings due to a reduction in disconnections, reconnections, and complaints. Other items such as less resource depletion, environmental benefits, job creation, improved housing stock, and participant comfort also must be valued.

Second, commissions and utilities may become involved with low-income weatherization for reasons of social equity. Due to the compelling need for low-income weatherization and the declining resources in the federal weatherization program, utilities and regulators may determine that participating in a partnership to help low-income customers is a legitimate role. Others may argue that it is inappropriate to shift a public problem to a private entity. Utilities may be willing to become involved as long as the burden of funding is not shifted exclusively to them. Commissions may mandate low-income services from an equity perspective because low-income households do not typically participate in other utility conservation programs due to a lack of financial resources.

Third, commissions and utilities may be willing to pursue low-income weatherization if utility arrearage reduction and improved bill payment behavior of low-income customers can be shown to result. Research must be undertaken to determine if there is a relationship between weatherization, arrearage reduction, and bill payment behavior.

Integration of Programs and Services

As more utilities implement low-income weatherization programs, integrating program services with state DOE-funded weatherization programs becomes an issue. Operating two independent programs can lead to duplication of services, excessive administrative costs for both programs, client confusion about which program to participate in, and incomplete weatherization services due to each program having limited resources. In addition, program constraints (such as per-job funding caps or the exclusion of certain energy conservation or health and safety measures) may limit the depth and quality of the weatherization work.

Integrating the two programs has the advantages of minimizing duplication and reducing program delivery and administrative costs. Integrated outreach, audit and assessment, construction management, installation, and inspection services can lead to a more cost-effective program that can better serve low-income people. Utilities have resources and can offer some program components that are not allowed by the DOE weatherization program or are not provided due to limited funding. These can include asbestos removal, heating system maintenance, energy education, electric weatherization (lighting and appliances), weatherization services to public housing, and the weatherization of buildings in the process of rehabilitation by community-based organizations or local governments.

The two programs can be integrated to varying degrees: (1) coordination between programs and referrals for program services; (2) cooperating on individual jobs, including cofunding or piggybacking measures; (3) agreeing on some common goals and objectives, such as eligibility, outreach procedures, and target population; or (4) complete program integration, including delivery systems, administration, and evaluation criteria.

Where feasible, other low-income energy services should be

integrated with weatherization programs. Packaging services allows for the most extensive menu of services for the customer. This can include problem identification and referral, low-income energy assistance, budget counseling, payment plans, and complete weatherization services. Integration of services is a must if low-income households are to gain control of their energy consumption. Without full energy services, low-income people will not be sufficiently empowered to take control of this component of their life.

Targeting of Services

Faced with the current situation of expanding need and limited resources, programs must target services and expenditures. Targeting services will help ensure that the greatest benefit is being achieved and that those households with the greatest need are being served.

Buildings that need weatherization can be screened and prioritized using energy consumption, income, energy burden, or arrearage level. For instance, targeting those buildings with the highest energy consumption will likely increase the mean energy savings resulting from the program. However, the program may not serve those households with the lowest income. If screening criteria are used, it is extremely important that they are developed carefully and in accordance with well-defined program goals. This technique works best if the weatherization program is coordinated or integrated with other low-income energy services, such as energy assistance, budget counseling, and payment plans. The integration of these programs helps to provide a wider range of services to low-income households and can significantly ease the burden of energy costs.

Low-Income Rental Issues

The client base for many low-income weatherization programs is shifting. In the early years of low-income weatherization, programs targeted single-family and one- to four-unit owner-occupied properties. In the last few years, the percentage of rental buildings being weatherized has steadily increased because outreach services are improving, programs that complement weatherization are being developed (such as budget counseling and payment plans), and, in some areas, many of the eligible single-family, owner-occupied units have been weatherized. Most households remaining to be weatherized are rental properties owned by nonresidents. These buildings, which can be small (one to four units) or large (five-plus units), present an entirely new set of issues to policy makers and weatherization providers.

The key weatherization policy issue in rental property occupied by low-income households is to ensure that the benefits of weatherization accrue to tenants and not to landlords. There are four potential areas where landlords can unduly benefit from weatherization. First, in individually metered buildings, landlords can raise the rent (because the building has been improved) and potentially capture some and perhaps all of the energy savings due to weatherization. Second, in master-metered buildings where energy costs are included in the rent, landlords might not decrease the rent to pass some or all of the energy savings to tenants. Third, landlords might evict low-income tenants and replace them with higher-income tenants after weatherization (units have added value due to lower energy costs and increased comfort). Fourth, landlords who sell their weatherized property might receive a higher selling price due to the weatherization improvements.

To prevent landlords from receiving the benefits of weatherization, the original DOE Weatherization Assistance Program (WAP) statute and regulations state that the benefits of weatherization must primarily accrue to tenants. It also states that rents cannot be raised solely due to weatherization and no undue or excessive enhancement can occur to the value of these properties. In 1989, DOE amended the WAP regulations to allow state providers to require landlord financial participation before rental properties receive weatherization in states with evidence of landlords capturing the benefits of weatherization. Rebates to tenants, extended rent freezes, and in-kind labor and materials are alternatives to financial contributions by landlords (DOE 1989a).

In a study of the impact of the DOE statute designed to prevent landlords from receiving the benefits of low-income weatherization, the National Consumer Law Center (NCLC) found that landlord agreements do not adequately ensure that the weatherization benefits are received by tenants. Agreements are not legally binding, do not specify which benefits should accrue to whom, and leave enforcement to the tenants, who have little authority to require compliance. Many states believe that if landlord agreements are specific, legally binding, and enforceable, landlords may not participate in the program. However, NCLC believes that legally binding landlord agreements are essential to protect against rent increases, evictions, and the sale of weatherized property. In master-metered buildings, energy savings must be provided to tenants via a reduction in rent. Tenants could become third-party beneficiaries to the landlord agreement, and thus could enforce the agreement and ensure future tenants are notified of their rights (NCLC 1988a).

Enforceable landlord agreements and landlord contributions pose significant problems for weatherization providers, to whom may fall the costly administrative burdens of tracking the benefits received by landlords. Depending on tenants to enforce the agreements seems untenable, since many are living in crisis, fear eviction, are highly mobile, and have little interest in tracking the benefits received by landlords. Agreements and contributions can discourage landlords from participating in programs, and thus may preclude the low-income household from receiving needed weatherization. Contribution levels that do not restrict participation are difficult to determine given the diversity of rental property ownership. Weatherization providers do not want to become collection agencies if landlords fail to pay their contributions. Some landlords will also be low-income people from whom a contribution would be a hardship. Providers might not meet production goals if agreements and contributions become detriments to participation by landlords.

Market research should be conducted to determine the extent of landlord benefits in each locale. Strategies could then be developed to more appropriately target the problem. Wisconsin Gas Company and the Wisconsin Department of Health and Social Services, the state administrators for the DOE program, are conducting a research study to determine the extent of landlord and tenant benefits from both the utility and state low-income programs in Milwaukee. The results will determine if program changes are necessary to ensure tenant benefits and minimize landlord enhancement (Wisconsin Gas Company 1990).

The final rental issue pertains to the weatherization of large rental properties (five-plus units) occupied by low-income households. These buildings are concentrated in large urban areas and contain heating systems more complex than those most weatherization providers are trained to service. Additional training of providers, or partnerships with organizations and contractors with large rental property experience, is critical in order to improve the energy efficiency of this building stock.

Technical Sophistication

While the technical sophistication of weatherization programs has increased dramatically, there is still plenty of room for improvement. Many of the technical improvements discussed in this chapter were implemented in pilot programs and field tests. The implementation of these improvements must become more widespread so that the state of the art becomes standard practice. Programs should continue many of the improvements begun during the past five years. Important steps are as follows:

- Routinely incorporate more comprehensive lists of effective energy conservation measures
- Use more accurate energy audits and more comprehensive measure selection protocols
- Target expenditures to households, buildings, and measures that provide the greatest opportunities for cost-effective conservation
- Incorporate energy education services
- Implement better installation procedures
- Evaluate and monitor the program to provide feedback on the performance of the program and identify areas for further improvement

The implementation of these improvements will result in a significant increase in program impact and cost-effectiveness nationwide.

Technical improvements should be developed and implemented regionally. What works well in one program with a certain housing stock and climate may not be effective in other regions. Regional approaches should be developed locally and supported by DOE and the utilities, particularly in cooling-dominated regions, where the installation of measures designed to reduce cooling energy use and the assessment of opportunities for saving cooling energy and demand are not widespread.

In addition to applying current technical improvements, programs should continue designing and developing innovations. Research on new or improved measures, installation procedures, audits and measure selection protocols, diagnostic tools, energy education, and evaluation methodologies should be encouraged, funded, and coordinated.

Energy Education

Energy education is an emerging component of low-income weatherization services. Households that have received weatherization have typically not known how to interact with a thermally efficient structure or understood the benefits of an energy-efficient lifestyle. As a result, some of the savings that were assumed to be captured by weatherization were lost due to occupant behavior. The rise in energy education programs can partially be attributed to changes in DOE rules that have allowed energy education as a program expense. In addition, many program operators have identified occupant education as an important part of the weatherization of a home. The availability of oil overcharge dollars has also spurred the development of these programs. Presently, 33 states have some form of energy education program (Quaid 1990).

Most energy education programs complement weatherization services. Program depth and delivery systems vary with the resources available. Workshops, home visits, feedback devices, videos, and written materials have all been components of education programs. Many programs have included incentives (money or energy-efficient devices) for participation and as a reward for attaining a set level of energy savings or for modifying specific behavior. Programs have targeted the installation of low-cost conservation measures such as caulking and weather stripping, attempted to modify energy use behavior (such as setting thermostats and appliance usage), and sought to determine strategies (budget plans and fuel assistance) to mitigate the impacts of utility bills (Quaid 1990).

The success of client/customer energy education programs has been difficult to gauge. Only recently have rigorous evaluations of these programs been performed to determine their effectiveness. Evaluation results have shown significant energy savings for these programs (Gregory 1989b; Witte and Kushler 1989; The Energy Collaborative 1989; Quaid 1990). Many programs that combine weatherization and energy education are experiencing additional weather-corrected space heating savings between 4% and 8%. Simple paybacks of less than three years are common for most of the programs (Quaid 1990).

Many issues must be resolved before energy education becomes a standard practice in weatherization programs. The major issue is the persistence of savings. Before funding, infrastructure, and other resources are expended to design and develop many full-scale programs, the longevity and reliability of savings should be determined. Evaluations performed to date have not analyzed savings in future years (Quaid 1990). With many programs incorporating incentives for customers to participate and reduce consumption, long-term impacts of the program without these incentives must be estimated. Methods of measuring program results must also improve. With a population subject to high attrition rates, the use of common billing analysis methodologies may be problematic due to data quality issues. Self-selection bias may also be a problem due to the voluntary nature of the program and the fact that initial programs may attract early adopters, people who are more likely to succeed in the program (Quaid 1990).

Programmatic issues must also be addressed. If funding for

weatherization dwindles in future years, new programs may be difficult to justify and sustain, especially if fewer homes are weatherized. Methods of integrating low-income services need to be explored, so energy education becomes an option even for a household not receiving weatherization services.

Finally, the question of resources has to be addressed. It is important to determine the appropriate delivery mechanism for energy education, the depth of services offered, and the population that may benefit the most from education. An infrastructure will have to be created to sustain education efforts that may divert resources from existing services.

These issues should not be insurmountable. Occupants are an important part of the weatherization services equation, and weatherization and energy education services must be balanced to optimize benefits obtainable by each segment of the low-income population. In this manner, program resources will be used most efficiently, and low-income households will be provided the services that help them most.

Impact on Program Delivery Systems

A decade ago the DOE Low-Income Weatherization Assistance Program was staffed as an employment and training program, providing CETA jobs to unskilled laborers. Recent advances in technical sophistication have forced managers to deal with new personnel issues. Many programs do not have enough trained and experienced people to implement the types of technical improvements discussed in this chapter. Pressures to implement changes too quickly without a reasonable plan usually reduce productivity. As utilities move to implement low-income programs, they often assume that DOE weatherization agencies can gear up to handle the increased workload. Programs in many parts of the country have severe shortages of trained and experienced field staff, and there are serious questions about the ability of these programs to incorporate the improvements discussed here. A program manager needs certain management and business skills to operate a weatherization program. Significantly more skills are needed to implement dramatic changes in a program and to balance DOE and utility weatherization activities.

A comprehensive training program aimed at enhancing the technical, communication, and business skills of existing and new program staff should be developed. As in other industries undergoing rapid change and expansion, this effort is needed to retool the approach to weatherization, incorporate what has been learned about buildings and energy use during the last ten years, and increase the skill level of program staff. The quality and quantity of existing training must be increased. Technical skills training should be experiential and conducted in the field wherever possible. Follow-up and reinforcement is critical to the success of technical training.

Program Evaluation and Monitoring

Evaluation and monitoring of low-income weatherization programs are critical activities that all programs should perform. A program manager would not think of running a program without monitoring the budget—why run a program without monitoring its performance? Evaluation and monitoring provide a feedback loop that allows program staff to learn from and about their own work, encourages self-evaluation, and creates an environment where staff are always looking for ways to improve the program. The end result is a more effective program. Evaluation has played a key role in Ohio, Michigan, Minnesota, Wisconsin, and other states where the performance of weatherization programs has improved dramatically. Evaluation is particularly important now as weatherization programs are pressured to prove their effectiveness.

Evaluation activities should be incorporated at two levels: routine formal evaluation and ongoing monitoring by management staff. Formal evaluation should be performed to document program effectiveness for funding sources, upper management, field staff, and colleagues. Ongoing monitoring should be performed as a routine management task, much like reviewing a monthly budget or balance sheet. Both types of evaluations should investigate energy and demand impacts, cost-effectiveness issues, and the process of the program. Future evaluations are likely to incorporate process evaluation (reviewing costs, productivity, delivery systems, and management approaches) as this area has a large potential for program improvement. Evaluation does not have to be prohibitively expensive. Well-planned and integrated evaluation activities add little to the cost of a program and provide significant benefits in future increases in program effectiveness.

The quality of evaluations must also improve in the near future. Where possible, evaluations should investigate the persistence of energy savings (Gregory 1989a). Impact and process evaluations should be integrated. Improved analysis methodologies should be developed. Better techniques to measure energy savings in homes with multiple fuels should be developed. Reports should include estimates of uncertainty and provide all pertinent information about the program and the results of the evaluation so that readers can interpret and learn from the evaluation. Some standardization may be required to enhance the quality of evaluations and the transferability of results (Schlegel and Pigg 1989).

DOE is currently conducting a three-year, comprehensive, national evaluation of the federal low-income weatherization program. The evaluation is designed to assess the energy savings, nonenergy impacts, and cost-effectiveness of the program. In addition, the study will analyze factors that may cause energy savings and cost-effectiveness to vary.

Increasing Energy Savings

The special energy-related problems of low-income people have been a matter of concern on the part of the federal and state governments and the utility industry since the early 1970s. Since that time, billions of dollars have been spent for low-income weatherization programs. From an energy savings perspective, these programs have not been as effective as anticipated. While annual energy savings have been increasing, further improvement is needed.

Two interrelated strategies to improve program impact are to increase the mean or median energy savings and to reduce the number of households that show negative savings. The technical improvements discussed in this chapter are designed to increase energy savings. More effective targeting of services can be used to invest program funds where there are greater opportunities for conservation, resulting in larger savings. Better assessment and installation procedures will ensure that the optimal package of measures for an individual building is recommended and installed, resulting in higher savings for the building and the program. The comprehensive investigation of unique situations in buildings will help ensure that some savings are achieved in all buildings, and that savings are maximized overall. Flexibility in program expenditures will allow programs to invest funds where there are opportunities for increased benefits and prevent spending money where there is little opportunity for conservation. Energy education should be incorporated so that occupant-directed conservation can be achieved and so that occupants fully understand the operation of newly installed measures.

The level of energy savings must be increased so that households recognize the differences in their energy bills. When people see conservation working, they are motivated further to conserve. Programs must have a greater impact if the burden of energy costs on low-income households is to be relieved.

Improving Cost-Effectiveness

Low-income weatherization programs must have a positive impact on the participant and be efficient in delivering services. Typically, low-income programs have been evaluated on the basis of energy savings. Recently, the issue of cost-effectiveness has received more emphasis. Determining cost-effectiveness raises the question: was the impact of the program worth the costs incurred in delivering the program? Managers and evaluators need to consider both the impact of a program and its cost-effectiveness when determining the performance of a low-income program.

To determine the cost-effectiveness of energy conservation measures, the value of the energy savings must exceed the installation and operation and maintenance costs. For the entire program to be cost-effective, the benefits must exceed the costs of the measures plus the program costs. The comparison of benefits and costs can be performed using a benefit/cost ratio, net present value, or simple payback calculation. It is important that cost-effectiveness be assessed by at least one of these methods and that benefits and costs be clearly identified. Since there is an ongoing debate on how to determine cost-effectiveness, others need detailed information so they can calculate cost-effectiveness in the manner with which they are most comfortable.

With the advent of utility least-cost planning, another dimension of cost-effectiveness must be addressed. Conservation and load management programs must meet certain economic tests before they can be pursued by a utility (Krause and Eto 1988). These tests typically look at the benefits of programs in terms of reduced electricity or gas supply costs and the associated external impacts such as reduced environmental impacts. Costs are viewed in terms of the utility cost of the program and the participants' cost of the program. Low-income programs, particularly natural gas programs, may not pass the economic tests due to the full cost of the program being paid by the utility and the relatively low marginal cost of energy faced by many utilities today. Economic cost-effectiveness criteria could limit many low-income programs to serving high-consumption customers with long-life conservation measures with paybacks of less than five years. This is particularly true for programs serving natural gas customers. Low-income programs will meet most societal tests if the external impacts of the program can be quantified sufficiently to be given a numeric value. These include arrearage reduction, improved bill payment behavior, costs saved from disconnection, reconnection and complaint handling, economic impacts from the development of a conservation delivery system such as job and business creation and less dollar flow from communities, and the mitigation of environmental impacts of energy production and delivery.

To date, little comprehensive analysis has been performed on low-income programs. Methodologies must be developed to place a numeric value on the benefits of the program that are not easily quantifiable. This step is critical, as development and implementation of utility low-income programs may depend on satisfying the economic tests associated with utility least-cost planning.

Conclusions

Low-income weatherization programs have matured dramatically in the last 15 years. They have always been programs of great promise, but that promise is only recently beginning to be realized. The programs have evolved technically and programmatically. Energy savings from weatherization, while still not at their theoretical potential, are improving with many programs reporting average annual savings of 15% to 20%. Several demonstration and pilot programs have achieved savings in the 20% to 30% range, indicating that still further improvement is possible. Increasing cost-effectiveness has become an objective for most programs. Program evaluation and ongoing monitoring are beginning to be valued as important management and feedback activities.

The program is at an important crossroads. Federal funding is beginning to decline while a large backlog of homes remains to be weatherized. Low-income people by and large have not benefitted from the economic recovery of the last decade and are not better off today than they were ten years ago. On a percentage basis, they still spend three to five times more of their income for energy than does the average American. Some utilities, on the other hand, face bad-debt write-offs in excess of 0.5% of net revenues due at least in part to nonpayment by low-income customers.

What is needed is a new will to go out and finish the job of low-income weatherization. The task can be completed within 15 years. Few programs can claim as many benefits, which include delivering conserved energy at costs that are competitive with some energy sources, cutting the energy bills of low-income people, reducing utility arrearages and associated costs, reducing the amount of carbon dioxide entering the atmosphere from fossil-fuel combustion, creating jobs, and improving the overall quality of the low-income housing stock.

Low-income weatherization programs could be a model for public-private cooperation. Government at all levels and the utility industry could work together to meet a common goal, through a standardized and integrated program, targeted to meet the energy needs of low-income people. Utilities and regulatory commissions are beginning to recognize the value of low-income weatherization. The government and utility sectors should initiate the necessary dialogue.

The low-income weatherization problem is not insurmountable. In fact, it is not even a particularly difficult problem. While it is important to continue to investigate further technological improvements, very adequate and effective tools have been developed and are available now. To borrow a phrase from a current advertisement, all we have to do is "Just Do It."

References

Bane, M., and D. Ellwood. 1989. "One Fifth of the Nation's Children: Why Are They Poor?" *Science* 245 (8 September): 1047–1053.

Banerjee, A., and M. Goldberg. 1985. *Evaluation of Utility Weatherization Programs in Wisconsin.* Department of Statistics. Madison: University of Wisconsin.

Barnett, N. 1988. Remarks before the National Association of Regulatory Commissioners, Subcommittee on Consumer Affairs on Behalf of the California Public Utilities Commission. San Francisco.

Bonneville Power Administration. 1986. *Issue Alert: A Housewarming for the Northwest.* Portland, Ore.: Bonneville Power Administration.

——. 1990. Big Savings from Small Sources: How Conservation Measures Up. Portland, Ore.: Bonneville Power Administration.

Cambridge Systematics. 1988. "Impact Evaluation of the Low-Income Direct Weatherization Program." Berkeley, Calif.: Cambridge Systematics.

Carmody, J. 1986. "Evaluation of the Low-Income Weatherization Program in Minnesota." Building Energy Research Center, University of Minnesota, Minneapolis.

Crenshaw, R., and R. Clark. 1982. *Optimal Weatherization of Low-Income Housing in the U. S.: A Research Demonstration Project.* Washington, D.C.: National Bureau of Standards.

DOE: See U.S. Department of Energy.

Economic Opportunity Research Institute. 1986. *Low-Income Energy Programs at Mid-Decade: Limits and Opportunities.* Washington, D.C.: Economic Opportunity Research Institute.

Ellison, S., C. Dunklee, J. Gatling, and B. Jacobsen. 1989. "An Experiment in Cooperation: The Connecticut Low-Income Weatherization Program (WRAP)." In *Demand-Side Management: Partnerships in Planning. Proceedings from the Electric Council of New England National Conference on Utility DSM Programs.* Vol. 1. EPRI CU-6598. Palo Alto, Calif.: Electric Power Research Institute.

The Energy Collaborative. 1989. *Partners in Energy Savings: Final Report and Evaluation.* Minneapolis: The Energy Collaborative.

Goldberg, M., A. Jaworski, and I. Tallis. 1984. *Low-Income Weatherization*

Program Study. Vol. 4, *Fuel and Household Analysis.* Madison: Wisconsin Division of State Energy, Department of Administration.

Gregory, J. 1987. *Ohio Home Weatherization Assistance Program. Final Report.* Columbus: Ohio Department of Development, Office of Weatherization.

——. 1989a. "Durability of Fuel Savings for Single-Family Households: Ohio Home Weatherization Assistance Program (HWAP), Program Years 1985 and 1986." In *Proceedings for the 1989 Conference on Energy Program Evaluation: Conservation and Resources Management.* Argonne, Ill.: Argonne National Laboratory.

——. 1989b. *Ohio Home Weatherization Assistance Program Fuel Savings Study.* Cleveland, Ohio: Center for Neighborhood Development, College of Urban Affairs, Cleveland State University.

Haeri, M., B. Bronfman, and D. Lerman. 1988. *Evaluation of the Bonneville Power Administration Low-Income Residential Weatherization Program.* Portland, Ore.: ERC International.

Hewitt, D., B. Senti, D. Salvesen, C. Gandehari, and L. Thiel. 1984. *Low-Income Weatherization Program Study.* Vol. 3, *Technical Findings.* Madison: Wisconsin Division of State Energy, Department of Administration.

Horowitz, M., and P. Degens. 1987. *Evaluation of the Utility Weatherization Assistance Program.* Portland, Ore.: International Energy Associates Limited.

Krause, F., and J. Eto. 1988. *Demand Side: Conceptual and Methodological Issues.* Vol. 2 of *Least-Cost Utility Planning Handbook for Public Utility Commissioners.* Washington, D.C.: National Association of Regulatory Utility Commissioners.

Kushler, M., P. Witte, and J. Stanley. 1987. *A Second Year Assessment of the Home Repair and Weatherization Component of the Michigan Energy Assurance Program.* Lansing: Michigan Public Service Commission.

Kushler, M., and P. Witte. 1988. "An Evaluation of the Fuel Savings Results of a New Weatherization Measures Priority System." Lansing: Michigan Department of Labor, Bureau of Community Services.

McBride, J., B. Castelli, D. Desmond, and M. Kelly. 1988. "The Critical Needs Weatherization Research Project." In *Proceedings from the ACEEE 1988 Summer Study on Energy Efficiency in Buildings.* Washington, D.C.: American Council for an Energy-Efficient Economy.

McCold, L., J. Schlegel, L. O'Leary, and D. Hewitt. 1988. *Field Test Evaluation of Conservation Retrofits of Low-Income, Single-Family Buildings in Wisconsin: Audit Field Test Implementation and Results.* ORNL/CON-228/P2. Oak Ridge, Tenn.: Oak Ridge National Laboratory.

Meridian Corporation. 1988. *Weatherization Evaluation Findings: A Comparative Analysis.* Alexandria, Va: Meridian Corporation.

National Center for Appropriate Technology. 1979. *Community Action Agency Appropriate Technology Needs Assessment.* Butte, Mont.: National Center for Appropriate Technology.

——. 1990. *National Community Action Agency Weatherization Program Survey.* Butte, Mont.: National Center for Appropriate Technology.
National Consumer Law Center. 1988a. *NCLC Analyzes Protections Accorded Tenants in Weatherized Units.* Washington, D.C.:. National Consumer Law Center.
——. 1988b. *National Utility Arrearage Survey Data.* Photocopy of unpublished study.
——. 1989. *Energy and the Poor—the Forgotten Crisis.* Washington, D.C.: National Consumer Law Center.
NCAT. See National Center for Appropriate Technology.
NCLC. See National Consumer Law Center.
N.Y. PSC. See State of New York Public Service Commission.
Penn. PUC. See Pennsylvania Public Utilities Commission.
Pennsylvania Public Utilities Commission. 1987. *Preliminary Evaluation of the Cost-Effectiveness of Residential Low-Income Usage Reduction Programs.* Harrisburg: Pennsylvania Public Utilities Commission.
PSC Ohio. See Public Service Commission of Ohio.
PSC Wis. See Public Service Commission of Wisconsin.
Public Service Commission of Ohio. 1985. *Opinion and Order on the Investigation into Long-Term Solutions Concerning Disconnection of Gas and Electric Service in Winter Emergencies.* Case No. 83-303-GE-COI (Phase II). Columbus: Public Service Commission of Ohio.
Public Service Commission of Wisconsin. 1982. *Investigation on the Commission's Own Motion to Reconsider Whether Class A Private Electric and Gas Utilities Should Provide Conservation Financing to their Customers to Promote Weatherization and What Suggestions Can Be Made to the Commission for Improving Conservation Programs for Natural Gas and Electric Utilities.* Docket No. 05-UI-12. Madison: Public Service Commission of Wisconsin.
Quaid, M. 1990. "Low-Income Energy Education Programs: A Review of Evaluation Results and Methods." In *Proceedings from the ACEEE 1990 Summer Study on Energy Efficiency in Buildings.* Vol. 7, *Government, Nonprofit, and Private Programs.* Washington, D.C.: American Council for an Energy-Efficient Economy.
Schlegel, J. 1990. "Blower Door Guidelines for Cost-Effective Air Sealing." *Home Energy* 7 (2): 34–38.
Schlegel, J., and S. Pigg. 1989. "Toward a Common Method of Reporting and Comparing Results from Evaluations of Residential Conservation Programs." In *Proceedings for the 1989 Conference on Energy Program Evaluation: Conservation and Resource Management.* Argonne, Ill.: Argonne National Laboratory.
——. 1990. "The Potential For Energy Savings and Cost-Effectiveness of Low-Income Weatherization Programs: A Summary of Recent Evaluations." In *Proceedings from the 1990 ACEEE Summer Study on Energy Efficiency in Buildings.* Vol. 6, *Program Evaluation.* Washington, D.C.: American Council for an Energy-Efficient Economy.

Sheladia Associates. 1988. *National Summary of State Plans*. Washington, D.C.: U.S. Department of Energy, Weatherization Assistance Program.

Shen, L., G. Meixel, T. Wilson, R. Belshe, G. Nelson, and M. Fagerson. 1986. *The State of Minnesota High-Level Weatherization Project*. Minneapolis: Minnesota Department of Energy and Economic Development.

Shen, L., G. Nelson, G. Dutt, B. Esposito, J. Fitzgerald, L. Gill, B. Hockinson, N. Hoffmann, and J. Shaten. 1990. *The M200 Enhanced Low-Income Weatherization Demonstration Project*. Minneapolis: University of Minnesota, Building Research Center.

State of New York Public Service Commission. 1989. *Order Instituting a Proceeding and Requesting Comments on a Motion of the Commission to Determine whether Major Gas and Combination Gas and Electric Utilities Subject to the Commission's Jurisdiction Should Establish and Implement a Low-Income Energy Efficiency Program*. Case No. 89-M-124. Albany: State of New York Public Service Commission.

Sweet, D., and K. Hexter. 1987. *Public Utilities and the Poor: Rights and Responsibilities*. New York: Praeger.

U.S. Department of Commerce. 1988. *Money Income and Poverty Status in the United States: 1987*. Washington, D.C.: U.S. Department of Commerce.

U.S. Department of Energy. 1989a. *Memorandum on Policy Guidance on Landlord Financial Participation in the Weatherization Assistance Program*. Washington, D.C.: U.S. Department of Energy.

——. 1989b. "Residential Energy Consumption Survey: Housing Characteristics 1987." Washington, D.C.: U.S. Department of Energy.

Vine, E., and I. Reyes. 1987. "Residential Energy Consumption and Expenditure Patterns of Low-Income Households in the United States." Berkeley: Lawrence Berkeley Laboratory.

Wisconsin Gas Co. 1990. "Request for Proposals to Determine the Benefits from Low-Income Rental Weatherization." Milwaukee: Wisconsin Gas Co.

Witte, P., and M. Kushler. 1989. "The Michigan Low-Income Weatherization Energy Education and Incentives Program." In *Proceedings from the 1989 Conference on Energy Program Evaluation: Conservation and Resources Management*. Energy Program Evaluation Conference, Chicago. Argonne: Argonne National Laboratory.

Chapter 7

U.S. Residential Appliance Energy Efficiency: Present Status and Future Policy Directions

Isaac Turiel, Douglas Berman, Peter Chan, Terry Chan, Jon Koomey, Benoit Lebot, Mark D. Levine, James E. McMahon, Greg Rosenquist, and **Steve Stoft**
Lawrence Berkeley Laboratory

Household appliances consume 35% of the electricity and 25% of the natural gas in the U.S. (EIA 1989). Therefore, energy-efficiency improvements in applicances for the residential sector can significantly impact power plant and local air emissions and, consequently, local air quality and global warming.

In this chapter, we discuss the recent history of the process for setting applicance energy-efficiency standards. Other policies for achieving energy savings from applicance efficiency improvements are also discussed (for example, providing financial incentives for the manufacture and purchase of more efficient models). The National Appliance Energy Conservation Act (NAECA), passed in 1987, amended the Energy Policy and Conservation Act (EPCA 1975) with respect to energy conservaton standards for applicances. Under NAECA, minimum efficiency standards were established for 11 consumer products. The U.S. Department of Energy (DOE) was instructed to periodically update these efficiency standards. In November of 1989, DOE completed its first appliance rulemaking, affecting refrigerators, freezers, and small gas furnaces. In mid-1991, standards for dishwashers, clothes washers, and dryers will be promulgated.

This chapter discusses issues related to the current standards and illustrates the standards' impacts on the model offerings of affected appliance manufacturers. For example, are the presently defined product classes too restrictive? How much research should be done in support of appliance efficiency rulemaking, and what are the roles of private industry and government? A number of technologies require additional research to determine whether they are commercially feasible: for example, evacuated panels for insulating refrigerators, microwave dryers, heat pump dryers, and dishwashing using lower-temperature water. Evaluating the benefits of some new design features, such as higher spin speeds, will require changing some DOE test procedures. At the end of this chapter, we discuss several alternative policy mechanisms to reduce appliance energy consumption.

Federal Standards

NAECA Standards

Under NAECA, manufacturers of certain residential appliances must meet legislatively specified energy conservation standards by certain dates. The energy efficiency or energy use for each product type is measured according to test procedures established by DOE.

Some inconsequential standards for clothes washers, clothes dryers, and dishwashers took effect on 1 January 1988. The requirements were: no pilot lights for gas clothes dryers, availability of a cold rinse option for clothes washers, and availability of an option to dry without heat for dishwashers. Most models of these products already satisfied these standards when NAECA was written.

The first standards of any consequence took effect on 1 January 1990. They apply to refrigerators, freezers, room air-conditioners, and water heaters. The NAECA legislation mandates periodic updates of these initial standards. Because of national concern about global warming, these updates have been the focus of much attention. Many members of Congress and other interested parties have realized the vast potential for reduction of fossil-fuel use (and concomitant reduction of CO_2 emissions) through more stringent energy-efficiency standards.

NAECA standards for the products mentioned above eliminate the most inefficient models. As will be discussed, however, additional energy savings are possible if more stringent standards are imposed or if financial incentives are developed to encourage the manufacture and purchase of more efficient models.

A substantial amount of energy could be saved merely by

promoting the purchase of the most efficient models already being manufactured. An example for clothes washers is to encourage the manufacture and purchase of horizontal axis machines rather than vertical axis machines. The horizontal axis machines consume much less water than the much more common (in the U.S.) vertical axis washers. In the U.S., vertical axis machines are all top loaders, whereas horizontal axis machines are front loaders. However, horizontal axis machines could be built as top loaders, as they are in some European countries.

Another option is to encourage new technologies such as evacuated panels for refrigerator insulation. In this technology, a partially evacuated panel is placed between the outer cabinet and the inner plastic liner, reducing the heat transfer from the outside air to the interior of the refrigerator.

Large energy savings would result from the implementation of this more effective insulation in refrigerators. National energy savings over the period 1993–2015 would reach 10.8 quads (out of a total of 41 quads) if evacuated panel insulation were utilized in refrigerator-freezers and freezers (DOE 1989a and 1989c), compared to an estimated savings of 5.2 quads for the mandated 1993 standards discussed below. An important aditional advantage of evacuated panel technology is the elimination of chlorofluorocarbons (CFCs) from refrigerator and freezer insulation.

Products Covered by NAECA Standards

Table 7-1 shows the 13 product types covered by EPCA as amended in NAECA. For each product type, there are subcategories, called

Table 7-1. Consumer products covered under NAECA.

Product Types
Refrigerators, Refrigerator-Freezers, and Freezers
Room Air-Conditioners
Central Air-Conditioners and Central Air-Conditioning Heat Pumps
Water Heaters
Furnaces
Dishwashers
Clothes Washers
Clothes Dryers
Direct Heating Equipment
Kitchen Ranges and Ovens
Pool Heaters
Television Sets
Fluorescent Lamp Ballasts

product classes, which differ in fuel type or some other characteristic affecting energy use. For example, gas and electric clothes dryers are in separate product classes.

Table 7-2 shows the relative importance in terms of energy use of regulated electric appliances. Estimates of average energy consumption were calculated for models produced in 1990 according to test procedures established by DOE. For several appliances, usage has decreased substantially, causing actual consumption to fall below consumption projected from the test procedure. Estimated consumption for electric water heaters (5,500 kWh/yr) includes most of the energy consumed by dishwashers and clothes washers, which consume most of their energy to heat water. Therefore, to avoid double counting when estimating household consumption, one should not add energy consumed by dishwashers and clothes washers to that consumed by water heaters. The percentages of U.S. households with such appliances reflect the significance of these product types to total U.S. residential electricity use.

Table 7-3 lists the 1990 and 1993 standards of allowable energy use for ten classes of one product type: refrigerators, refrigerator-freezers, and freezers. Since energy use is a function of volume, the standards are in the form of equations that have volume as an independent variable. An example below shows how the energy standard

Table 7-2. Typical energy use of new electric appliances.

Product Type	kWh/yr+	% Households*
18-ft^3 Refrigerator-Freezer	950	100
15-ft^3 Upright Manual Freezer	700	42
Standard Dishwasher	920	50
Top-loading Standard Clothes Washer	1,200	72
Standard Clothes Dryer	1,050	45
40-gallon Water Heater	5,500	47
8,500 Btu/hr Room Air-Conditioner	700**	25
Central Air-Conditioners	3,000***	40
Heat Pumps	3,000***	16
19" Color Television	200	97

+ Estimates made according to DOE test procedures. Actual field use will differ.
* Source for % households with appliance type is *Appliance Magazine*, September 1989.
** Assumed hours of operation are 750.
*** Assumed hours of operation are 1,000.

Note: Average U.S. household total electric energy use equals 8,939 kWh/yr (DOE/EIA 1989). All electricity-using appliances are included.

for one product class—top-mount automatic-defrost refrigerator-freezers—is calculated. For this product class (accounting for almost 70% of annual refrigerator and refrigerator-freezer sales in the U.S.) the current maximum allowable energy use is given by

$$E = 23.5\ AV + 471$$

where AV, adjusted volume, equals refrigerator volume plus 1.63 times the freezer volume. For a typical 18-ft^3 model, the adjusted volume equals 21 ft^3 (fresh food and freezer volumes of 13.5ft^3 and 4.5 ft^3 respectively) and the allowable energy consumption is 965 kWh/yr. Appendix A lists the NAECA standards for some other product types.

Procedures for Setting Appliance Standards

Any new or amended standard must achieve the maximum improvement in energy efficiency that DOE determines is technologically feasible and economically justified. In determining whether a standard is economically justified, DOE decides whether the benefits exceed the burdens by considering seven factors specified by law:

1. The economic impact of the standard on the manufacturers and on the consumers of the products.
2. Savings in operating costs throughout the estimated average life of the product compared to the increase in product price (including installation and maintenance).

Table 7-3. Maximum allowable energy consumption for refrigerators, refrigerator-freezers, and freezers (kWh/yr).

Product Class	1990 Standards	1993 Standards
Manual Defrost Refr. and Refr.-Freezer	16.3AV⁺ + 316	13.5AV + 299
Partial Auto-Defrost Refr.-Freezer	21.8AV + 429	10.4AV + 398
Top-Mount Auto-Defrost Refr.-Freezer	23.5AV + 471	16.0AV + 355
Side-Mount Auto-Defrost Refr.-Freezer	27.7AV + 488	11.8AV + 501
Bottom-Mount A-D Refr.-Freezer	27.7AV + 488	16.5AV + 367
Top-Mount A-D with TTD* Features	26.4AV + 535	17.6AV + 391
Side-Mount A-D with TTD Features	30.9AV + 547	16.3AV + 527
Upright Manual Defrost Freezer	10.9AV + 422	10.3AV + 264
Upright Auto Defrost Freezer	16.0AV + 623	14.9AV + 391
Chest Freezers	14.8AV + 223	11.0AV + 160

+ AV means adjusted volume; for refrigerator-freezers, AV = Refrigerator Volume plus 1.63 times the Freezer Volume. For freezers, AV = 1.73 times Freezer Volume.

* TTD stands for through-the-door features such as ice makers.

3. Total amount of projected energy savings.
4. Any lessening of the utility of the product.
5. Impact of any lessening of competition.
6. The need for national energy conservation.
7. Other factors DOE considers relevant.

If DOE finds that the additional cost to the consumer of purchasing a product meeting an energy conservation standard level will be less than three times the value of the estimated energy savings during the first year, as calculated under the appropriate test procedure, there is a rebuttable presumption that the level set by the standard is economically justified. This criterion is equivalent to requiring a simple-payback period of less than three years. Thus, unless other factors take precedence, energy-efficient designs with payback periods of less than three years will be incorporated into new standards, but this policy does not preclude efficiency improvements with paybacks greater than three years from being incorporated into new standards.

The first step in the DOE rulemaking process is the publication of an Advance Notice of Proposed Rulemaking (ANOPR). The purpose of this notice is to inform all interested parties of the product types and classes for which DOE intends to consider setting standards. The ANOPR also sets forth the designs to be analyzed and the computer models to be utilized. Information received by DOE during the public comment period is considered in the preparation of the Notice of Proposed Rulemaking (NOPR). The NOPR presents the proposed policy, the results of the analysis, and the alternatives considered. During another public comment period, hearings are held in Washington, D.C. The oral and written comments received on the NOPR are considered in preparing a final rulemaking, which contains any new energy conservation standards.

Recent Updates of Standards for Refrigerators and Freezers

On 17 November 1989, the DOE published a notice in the Federal Register listing new energy-efficiency standards for ten classes of refrigerators and freezers (DOE 1989). Table 7-3 shows the standards that will take effect on 1 January 1993. These are the first energy-efficiency standards promulgated by DOE. Figure 7-1 shows the 1990 and 1993 NAECA standards and the annual energy consumption for top-mount auto-defrost refrigerator-freezers listed in the 1989 Directory of Certified Refrigerators and Freezers published by the Association of Home Appliance Manufacturers (AHAM

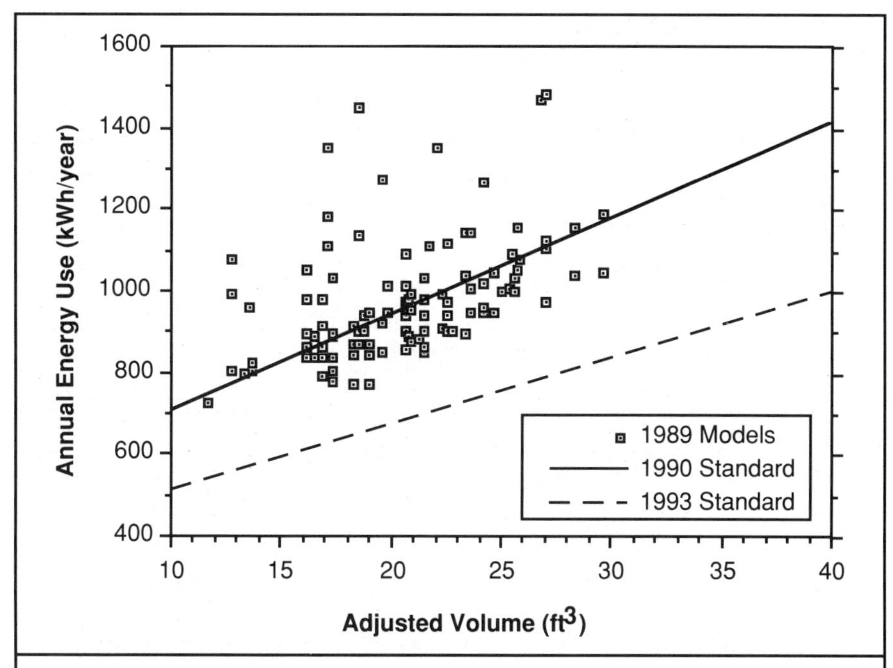

Figure 7-1. Energy consumption as a function of adjusted volume for top-mount auto-defrost refrigerator-freezer.

1989). As can be seen, the 1993 standards are significantly more stringent than the 1990 standards; in 1989, energy consumption for the same adjusted volume varied widely, and many inefficient models could no longer be manufactured after 1 January 1990. At the present time, no models for this class listed in the AHAM Directory meet the 1993 standards: all models must be redesigned (AHAM 1990). For all classes of refrigerators, refrigerator-freezers, and freezers, only 7 out of 2,114 meet the 1993 standards.

The 1993 standards are based on technical and economic analyses carried out for DOE by Lawrence Berkeley Laboratory (DOE 1989c). The analyses commence with a technical analysis of manufacturing costs required to improve the efficiency of each product class (called the engineering analysis). A manufacturer impact model (MIM) is used to generate consumer prices from manufacturer costs (DOE 1988). Life-cycle costs and simple-payback periods are calculated from the price and efficiency data. Additional maintenance and installation costs are also considered at this time. A residential end-use model (LBLREM) is used to calculate national energy savings from several trial standard levels (McMahon 1987).

Engineering Analysis

The engineering analysis provides information on efficiencies, manufacturing costs, and other appliance characteristics; this information is used in other components of the analysis. Different appliance features useful to the consumer are incorporated into the analysis through the creation of product classes, the subsets of appliance types. The engineering analysis develops manufacturer cost and efficiency data for a set of design options within each product class. The analysis is performed in the following steps: (1) select product class, (2) select baseline units, (3) select design options within each class, (4) determine maximum technically feasible designs, (5) develop and run energy-use models, (6) develop cost estimates, and (7) rank design options to generate cost-efficiency relationships.

To use the analysis of refrigerators and freezers as an example, for each product class, a baseline unit was selected with an energy use approximately equal to the maximum energy use allowed by the 1990 NAECA standard for refrigerators; then design options were developed for that unit. Table 7-4 shows these options for the product class of top-mount auto-defrost refrigerator-freezers. A simulation program was used to calculate energy consumption for

Table 7-4. Design options for top-mount auto-defrost refrigerator-freezers.

Foam Insulation Substitution in Cabinets or Doors
Increased Insulation Thickness in Cabinets or Doors
Reduced Heat Load of Through-the-Door Features
Double Door Gaskets
Improved Foam Insulation
Evacuated Insulation Panels
High-Efficiency Compressor Substitution
Adaptive Defrost
Fan and Fan Motor Efficiency Improvement
Anti-Sweat Heater Switch
Condenser Gas Heating
Increased Evaporator Surface Area
Hybrid Evaporator
Enhanced Heat Transfer Surfaces
Mixed Refrigerants
Improved Expansion Valve
Fluid Control Valve
Two-Compressor System
Variable-Speed Compressor
Two-Stage Two-Evaporator System
Use of Natural Convection Currents
Relocation of Compressor, Condenser, and Evaporator Fan Motor

each set of design options (Arthur D. Little 1982). This steady-state model calculates heat leakage into a cabinet and then determines the energy needed by the refrigeration system to maintain cabinet temperatures as specified in the DOE test procedure. Detailed information on the cabinet dimensions, insulation levels, compressor performance, heat exchanger effectiveness, and auxiliary equipment is needed to run the simulation model. These data were also used to calibrate the baseline unit to the measured energy use of an actual model.

Table 7-5 (pages 208 and 209) shows the results of running the simulation model for an 18-ft^3 top-mount auto-defrost refrigerator-freezer; the table shows percentage of compressor run time, cabinet heat gain, compressor power, auxiliary power, and daily and yearly energy use. Presently, CFC-11 is used for insulation and CFC-12 for the refrigerant in residential refrigerators and freezers. Both of these compounds break down under ultraviolet radiation in the stratosphere. The resulting chlorine-containing compound reacts with ozone, breaking it into an oxygen molecule and oxygen atom. For these simulations, it is assumed that either CFC-12 or a non-CFC refrigerant of efficiency equal to CFC-12's is available and that a non-CFC foam is used in the insulation. The cost of the CFC-11 substitute is estimated to be 2.5 times greater than the present cost of CFC-11; the cost of a CFC-12 substitute is estimated to be 2.5 times that of CFC-12. The resulting baseline manufacturer cost is approximately $4 higher for a non-CFC 18-ft^3 refrigerator than for one with CFCs now available. Simulations were also performed for a scenario using CFC-11 and CFC-12 (DOE 1988).

Most of the design options are self-explanatory. Actual data on compressor performance were obtained from compressor and refrigerator manufacturers. The foam insulation thermal conductance value for the baseline model is approximately R-8 per inch, or 0.127 Btu per inch of thickness per square foot of area per hour per number of degrees Fahrenheit difference across the cabinet shell (Btu-in/ft^2-hr-°F). This value assumes the use of a non-CFC foam such as HCFC-123 or HCFC-141-b, and is 2% higher than the value for a CFC-11–based foam. The fan motor power is chosen to be 8 W for the more efficient evaporator and condenser fan motors. After more efficient fans, the choice of the next design option branches: either increased side insulation or evacuated panels are chosen.

The evacuated panel thermal conductance value of R-18 per inch (0.055 Btu inch/ft^2-hr-°F) is assumed to apply to the walls of the cabinet and not to the doors. It is a composite of foam insulation (for structural rigidity) and evacuated panel. Any of three technologies

Table 7-5. Energy use of an 18-ft.^{3} top-mount auto-defrost refrigerator-freezer.

Level	Description	% Run Time	QS (Btu/hr)	PCOMP (W)	Auxiliary Power (W)		kWh Used per Day	Average Yearly Energy Use (kWh)
					Fans	Antisweat Heaters		
0	Baseline‡	0.45 0.42	264 264	186 187	23.5 23.5	19.0 5.0	2.88 2.35	955
1	Baseline plus Enhanced Evaporator	0.43 0.40	264 264	191 192	23.5 23.5	19.0 5.0	2.83 2.30	936
2	Level 1 plus Door Foam Insulation	0.41 0.37	241 241	191 192	23.5 23.5	19.0 5.0	2.67 2.14	878
3	Level 2 plus 5.05 EER Compressor	0.41 0.38	241 241	161 161	23.5 23.5	19.0 5.0	2.41 1.90	787
4	Level 3 plus 2" Door Insulation	0.40 0.36	230 230	161 161	23.5 23.5	19.0 5.0	2.35 1.83	763
5	Level 4 plus More Efficient Evaporator and Condenser Fans	0.40 0.36	230 230	160 161	16.0 16.0	19.0 5.0	2.26 1.75	732
6	Level 5 plus 2.60"/2.30" Side Insulation and 2.6" Back Insulation	0.38 0.34	218 218	160 161	16.0 16.0	19.0 5.0	2.19 1.68	706
7	Level 5 plus 3.0"/2.70" Side Insulation and 3.0" Back Insulation	0.37 0.33	210 210	160 161	16.0 16.0	19.0 5.0	2.14 1.64	690
8	Level 5 plus Evacuated Panel (K=0.055)	0.30 0.27	159 159	160 161	16.0 16.0	19.0 5.0	1.83 1.33	577

(continued on next page)

Table 7-5 *(continued)*.

Level	Description	% Run Time	QS (Btu/hr)	PCOMP (W)	Auxiliary Power (W)		kWh Used per Day	Average Yearly Energy Use (kWh)
					Fans	Antisweat Heaters		
9	Level 8 plus	NA*	159	NA	16.0	19.0	1.61	508
	Two-Compressor System	NA	159	NA	16.0	5.0	1.17	
10	Level 9 plus	NA	159	NA	16.0	19.0	1.55	490
	Adaptive Defrost	NA	159	NA	16.0	5.0	1.13	

‡18.0-ft^3 (20.8-ft^3 adjusted volume) refrigerator-freezer, side wall insulated with 2.20" foam in freezer, 1.9" foam in refrigerator, door insulated with 1.5" foam in freezer and 1.5" fiberglass in refrigerator, back insulated with 2.2" foam. Features include improved thermal seal gasket, antisweat switch, 4.5 EER compressor, bottom-mounted condenser, auto-defrost timer, 10W evaporator, and 13.5W condenser fans.

*Energy savings for these design options were calculated outside the simulation model.

for evacuated panels could be used to achieve this thermal conductivity value. These approaches are compact vacuum, aerogel, and powder-filled evacuated panels. The first approach creates a hard vacuum between rigid steel plates with glass spacers (Potter 1989). A 0.25-inch test panel has been produced with an R-10 (R-40 per inch) insulation value. Silica aerogel is a low-density form of porous silica glass. Measurements show that R-20 per inch is presently achievable at a pressure of about 75 torr (760 torr equal one atomosphere of pressure) (Quantum Optics 1989). The last approach, powder-filled evacuated panels, has progressed furthest. General Electric has produced limited quantities of these panels for installation in about 1,000 refrigerator-freezers (GEA 1989). A 0.05 thermal conductivity or R-20/inch was achieved at a pressure of about 20 torr. Because of unsatisfactory and inconsistent results, General Electric has withdrawn evacuated panel insulation from currently manufactured refrigerator-freezers.

A two-compressor system employs two separate evaporators and two compressors; one for the refrigerator and one for the freezer. This measure allows for a higher evaporator temperature in the refrigerator and a higher COP than for the freezer-refrigerator system. For adaptive defrost, defrosting is done when needed rather than through a timer control. The maximum technically feasible reduction in energy use occurs at level 10 in Table 7-5. Energy consumption is reduced by almost 50% from baseline, from 955 to 490 kWh/yr. This design incorporates two inches of foam insulation in the doors, a 5.05 EER (Energy Efficiency Rating) compressor, evacuated panels in the walls, a two-compressor system, and adaptive defrost.

Manufacturer Cost Data

Table 7-6 combines data on energy use and on manufacturer cost. The design options are ordered so that those that are relatively more cost-effective (a higher ratio of energy savings per dollar of manufacturer cost) are listed first. The manufacturer cost is the cost to the manufacturer of producing models with the design options shown and does not include markups to wholesalers or retailers. The cost data are mostly obtained from manufacturers and averaged in order to protect the confidentiality of data received from individual manufacturers. Independent estimates of the cost of purchased parts were obtained from compressor and fan motor manufacturers. Disaggregated manufacturer costs (materials, purchased parts, labor, tooling, and shipping/packaging) are found in Appendix A of the Technical Support Document (DOE 1989c).

Table 7-6. Manufacturer cost and energy consumption for a top-mount auto-defrost refrigerator-freezer (no CFCs).

Design	Design Option	Energy Use (kWh/yr)	Manufacturer Cost (1987$)
0	Baseline	955	224
1	0 + Enhanced Evapor. Heat Transfer	936	224.1
2	1 + Foam Refrigerator Door	878	225.55
3	2 + 5.05 Compressor	787	228.95
4	3 + 2" Doors	763	232.65
5	4 + Efficient Fans	732	241.65
6	5 + 2.6"/2.3" Side and 2.6" Back Insul.	706	249.10
7	5 + 3.0"/2.7" Side and 3.0" Back Insul.	690	253.95
8	5 + Evacuated Panels	577	287.65
9	8 + Two-Compressor System	508	337.65
10	9 + Adaptive Defrost	490	353.65

Volume = 18.0 ft^3, AV = 20.8 ft^3. Baseline includes 4.5 EER compressor, side wall insulation of 2.2" foam in freezer and 1.9" foam in refrigerator, and door insulation of 1.5" foam in freezer and 1.5" fiberglass in refrigerator. Design options are ranked in order of cost-effectiveness, with option 1 being the most cost-effective.

Simple-Payback Period and Life-Cycle Cost Analyses

Table 7-7 (next page) shows incremental and cumulative simple-payback periods and consumer prices for each design option for top-mount auto-defrost refrigerator-freezers, assuming the price of electricity in 1993 to be $.075/kWh in 1987 dollars. The cumulative paybacks were computed relative to design level 2 (foam door), the base case efficiency level predicted by LBLREM for 1993 new top-mount auto-defrost refrigerator-freezers. We chose this baseline for the cumulative payback period calculation to take into account projected efficiency improvements in the marketplace beyond the 1990 minimum efficiency standard. Therefore, cumulative paybacks are not applicable (NA) to the first two design options. The trial standard levels, shown in the first column, are potential standard levels DOE chose for further analysis. Trial level 2, with a cumulative payback period of 4.2 years, was selected as the final standard for this product class. This choice was based on all of the decision-making criteria discussed earlier.

Figure 7-2 (page 213) shows the life-cycle cost of purchasing (in 1993) and operating an 18-ft^3 top-mount auto-defrost refrigerator-freezer as a function of annual energy use. The life-cycle cost is computed for three discount rates (5%, 7%, and 10% real) and for an average equipment lifetime of 19 years. The projected energy price

Table 7-7. Payback periods of design options (years) for a top-mount auto-defrost refrigerator-freezer (no CFCs).

Trial Std.	Design Level	Design Options	Price (1987$)	Increm. SPP*	Cumul. SPP**
	0	Baseline	521.7	NA	NA
	1	0 + Enhanced Evapor. Heat Transfer	521.9	0.2	NA
	2	1 + Foam Refrigerator Door	525.1	0.7	NA
1	3	2 + 5.05 Compressor	532.3	1.0	1.0
	4	3 + 2" Doors	540.6	4.4	1.7
2	5	4 + Efficient Fans	559.5	7.7	3.0
	6	5 + 2.6"/2.3" Side Insulation	577.0	8.5	3.8
3	7	5 + 3.0"/2.7" Side Insulation	587.1	8.3	4.2
4	8	5 + Evacuated Panels	656.2	7.9	5.5
	9	8 + Two-Compressor System	760.7	19.2	8.1
5	10	9 + Adaptive Defrost	794.0	23.4	8.8

* Increm. SPP is the incremental simple-payback period. Incremental SPP is calculated by dividing the consumer price difference (between consecutive designs) by the dollar value of energy saved each year.
** Cumul. SPP is the cumulative simple-payback period. Cumulative SPP is calculated by dividing the consumer price difference (between the projected baseline for 1993 and the design under consideration) by the dollar value of energy saved each year.

of $.075/kWh is from *Annual Energy Outlook* (DOE/EIA 1989). The minimum life-cycle cost occurs at design level 8, evacuated panel insulation.

National Energy Savings

The LBLREM model projects a number of economic and energy-use variables that are employed to assess the impact of proposed standards on consumers, electric utilities, and appliance manufacturers. This section presents the model's energy-use projections assuming various trial standard levels.

Table 7-8 presents projections of total energy consumption of residential refrigerators, refrigerator-freezers, and freezers. In the base case, annual energy consumption for refrigerators, refrigerator-freezers, and freezers declines from 1.85 quads (162,000 GWh) in 1995 to 1.72 quads in 2000, then rises to 1.90 quads in 2015. The initial decline is due partly to efficiency improvements mandated by NAECA and partly to the naturally occurring replacement of older, less-efficient models with more efficient ones. The subsequent rise is due to an increasing number of households owning these products. A similar pattern is shown for standards levels 1 through 4 at lower energy consumption levels. For standards level 5,

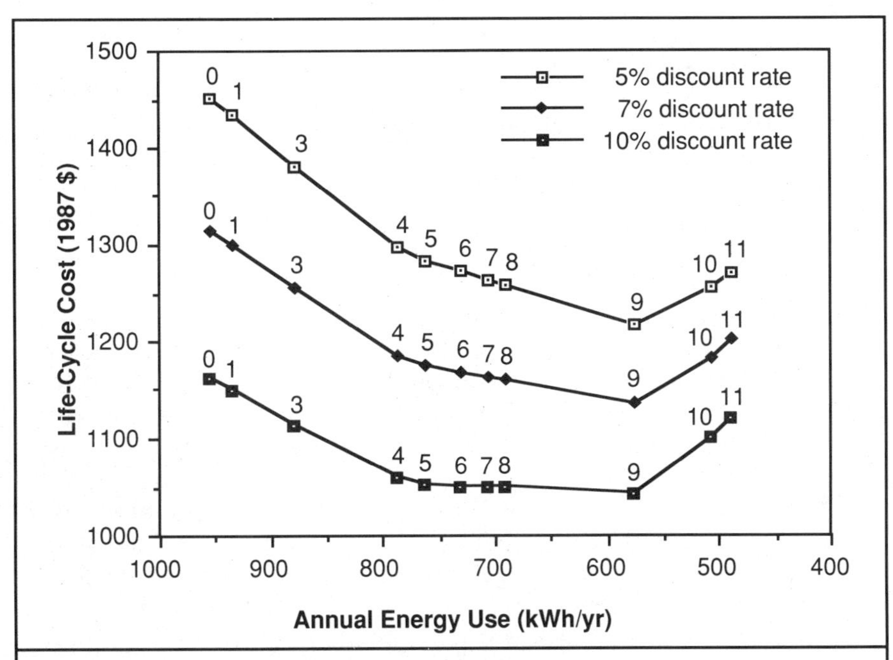

Figure 7-2. Life-cycle costs for design options of top-mount auto-defrost refrigerator-freezer without through-the-door features.

Table 7-8. Annual residential energy consumption by refrigerators, refrigerator-freezers, and freezers (in quads, or 10^{15} Btu).

		Standards Level[a]				
Year	Base	1	2	3	4	5
1995	1.85	1.83	1.82	1.80	1.77	1.75
2000	1.72	1.65	1.61	1.57	1.47	1.41
2005	1.72	1.60	1.54	1.46	1.30	1.20
2010	1.79	1.63	1.54	1.45	1.22	1.07
2015	1.90	1.71	1.61	1.50	1.23	1.06
1993-2015	41.1	38.7	37.4	35.9	32.5	30.3
Change from Base Case	–	−2.4	−3.7	−5.2	−8.6	−10.8
Cumulative Energy Savings	–	5.8%	9.0%	12.7%	21.0%	26.3%

a. Standards levels correspond to design levels in Table 7-7.

the improvement in energy efficiency is the dominant factor, resulting in a continuous decline in energy consumption through 2015.

The cumulative energy savings from standards over the period 1993–2015 range from 2.4 quads for standards level 1 to 10.8 quads for standards level 5; these savings correspond to 5.8–26.3% of base case electricity use for these appliances.

Updates of Standards for Dishwashers, Clothes Washers, and Dryers

In 1991, DOE will publish amended energy conservation standards for dishwashers, clothes washers, and clothes dryers. The standards will take effect three years after publication of the final rule. In Figures 7-3 and 7-4 (pages 215 and 216), we can see the range of efficiencies presently available in the marketplace for dishwashers and clothes washers. Manufacturers are not required to provide efficiency data for clothes dryers. Dishwashers have approximately a 20% range in efficiency from the most to least efficient models. For clothes washers the range is much greater, but much of this variation is due to the impact of having, or not having, warm rinse. The efficiency of a clothes washer is defined as the capacity in ft^3 divided by the energy use in kWh per cycle.

Engineering Analysis

Tables 7-9 through 7-11 show the energy-use data for dishwashers, clothes washers, and clothes dryers (manufacturer cost data cannot be presented at this time since final standards for these products have not yet been published). For the first two products, most energy use is for heating water. For dishwashers, the first design option, reduced water use, produces the greatest reduction in hot water use and thus energy consumption. This reduction can be accomplished by more efficient food filters, use of alternate spray arms (AEG 1989), or other methods. Other design options are improved motor efficiency, improved fill control, addition of a booster heater to the standard dishwasher, and reduced booster use for the water heating dishwasher. In Table 7-9 (page 217), annual energy use is calculated with recent usage data (Procter & Gamble 1989). The new usage figure of 229 cycles per year is substantially lower than the existing DOE test procedure value of 322 cycles per year.

For clothes washers, two design options can produce large energy savings: elimination of warm rinse and the use of horizontal axis machines. Elimination of the warm rinse option will reduce energy use by about 25%. A horizontal axis design will decrease hot water use by 64% and total water consumption by 31% (White

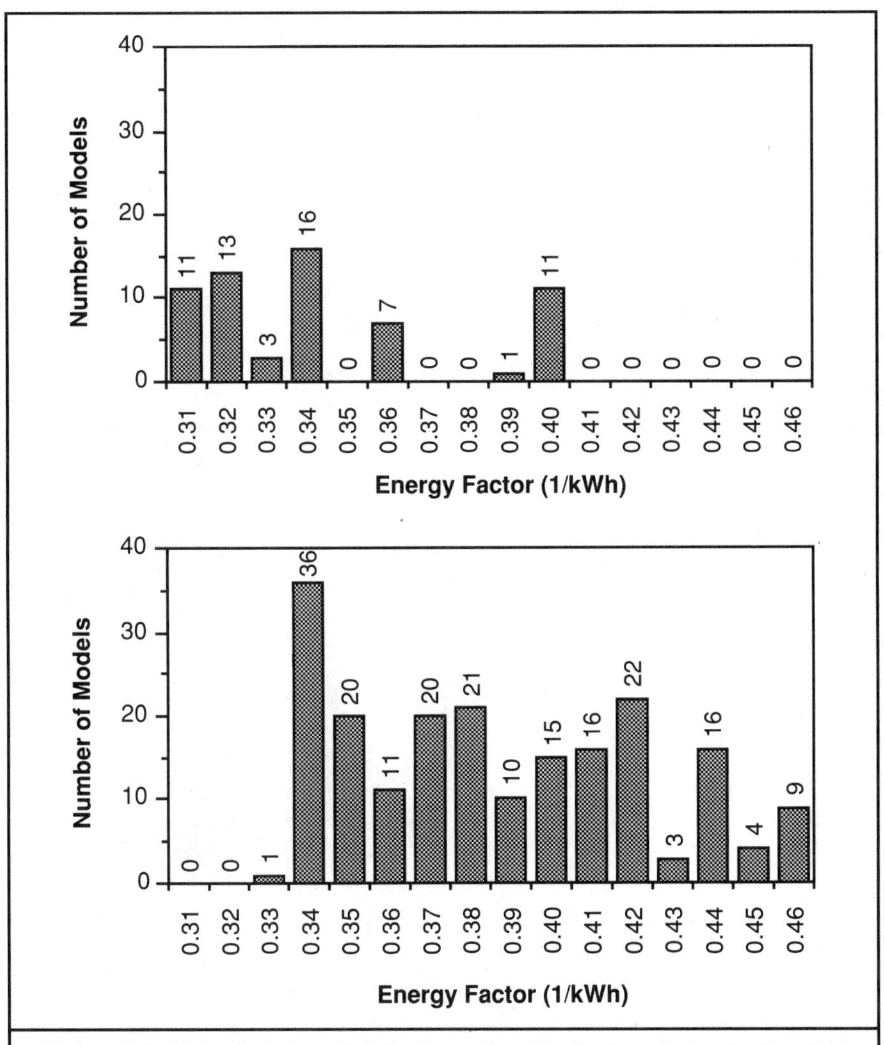

Figure 7-3. Energy factor distributions for standard and standard water-heating dishwashers.

Consolidated 1989) because the drum does not have to be filled as high as it does for vertical axis washers. In this design, the drum rotates about a horizontal rather than a vertical axis; the machine remains a top loader (Lebot 1990). Horizontal axis machines are exclusively used in Europe, where both top- and front-loaders are manufactured. Other design options, such as improved motor efficiency, thermostatically controlled mixing valve, and use of a

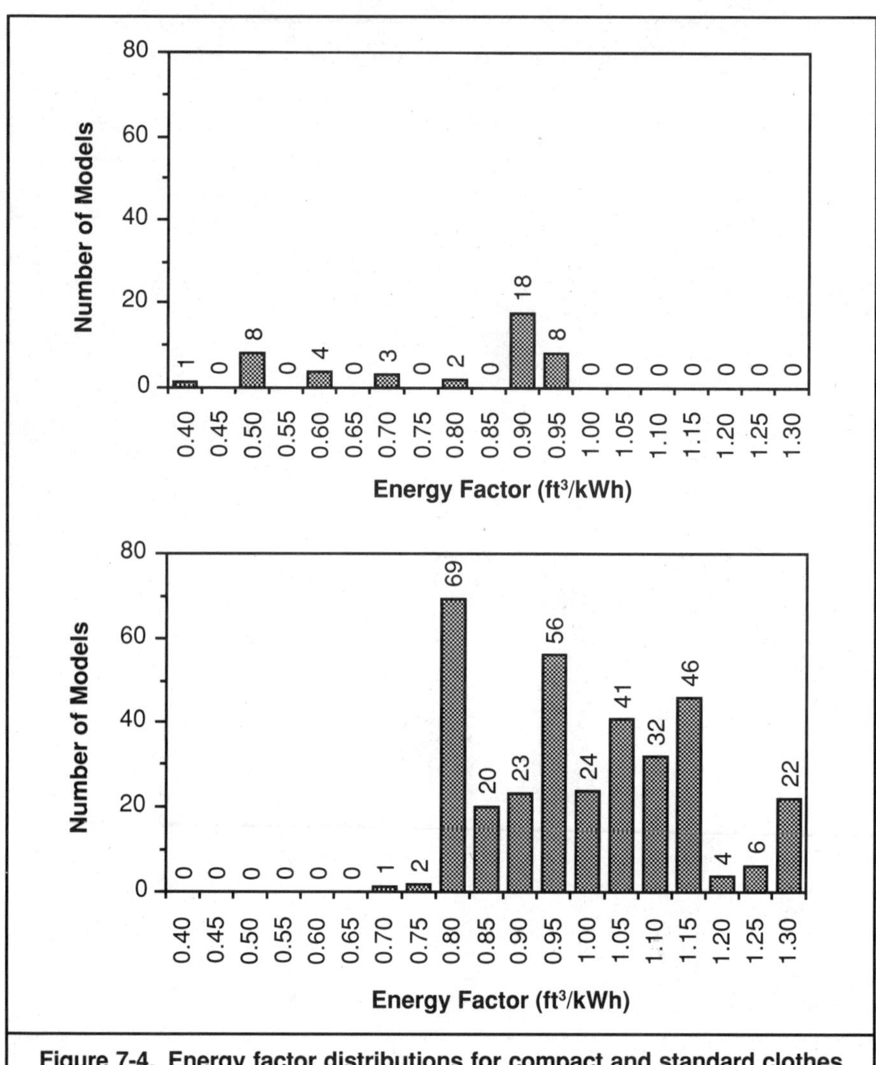

Figure 7-4. Energy factor distributions for compact and standard clothes washers.

plastic tub produce very small energy savings. Table 7-10 (page 218) assumes a usage factor of 380 cycles per year rather than the existing DOE test procedure value of 416 cycles per year (Procter & Gamble 1989).

In Table 7-11 (page 219), annual energy use for clothes dryers is calculated with a new usage factor of 359 cycles per year rather than the existing DOE test procedure value of 416 cycles per year. For

Table 7-9. Energy use data, standard dishwasher, 115V, Procter & Gamble 1988/98 data.

Level	Design Option	Energy Factor load/kWh	Energy Use[c] kWh/year	Motor Energy kWh/cycle	Booster Energy kWh/cycle	Dryer Energy kWh/cycle	Hot Water Energy kWh/cycle	Hot Water Cons. gal./cycle
0	Baseline	0.32	716	0.30	0.00	0.25	2.58	11.9
1	Reduce Water Use	0.38	598	0.30	0.00	0.25	2.06	9.5
2	1 + Improved Motor	0.39	585	0.24	0.00	0.25	2.06	9.5
3	2 + Booster Heater[a]	0.44	522	0.24	0.18	0.25	1.60	9.5
4	3 + Fill Control	0.45	510	0.24	0.18	0.25	1.55	9.3
5	1 + Booster Heater[b]	0.45	514	0.30	0.09	0.25	1.60	9.5
6	5 + Improved Motor	0.46	501	0.24	0.09	0.25	1.60	9.5
7	6 + Fill Control	0.47	490	0.24	0.09	0.25	1.55	9.3

a. Add Booster heater and heat 40% of hot water from 120°F to 140°F
b. Add Booster heater and heat 20% of hot water from 120°F to 140°F
c. 229 Cycles/year

Table 7-10. Energy use data, clothes washer, top-loading, standard capacity (2.6-ft^3), Procter & Gamble 1988/98 data.

Level	Design Option	Energy Factor ft^3/kWh	Energy Use[a] kWh/year	Motor Energy kWh/cycle	Hot Water Energy kWh/cycle	Hot Water Cons. Gal/cycle	Total Water Cons. Gal/cycle
0	Baseline 5 Setttings	1.04	950	0.27	2.23	10.3	33.3
1	Eliminate Warm/Warm Set.	1.32	746	0.27	1.70	7.9	33.3
2	Eliminate Warm Rinse	1.47	674	0.27	1.50	7.0	33.3
3	2 + Improved Motor	1.50	660	0.23	1.50	7.0	33.3
4	3 + Plastic Tub	1.50	657	0.23	1.50	6.9	33.3
5	4 +Therm. Mix. Valve	1.51	653	0.23	1.49	6.9	33.3
6	2 + Horizontal Axis	3.77	262	0.15	0.54	2.5	22.8
7	6 + Plastic Tub	3.79	260	0.15	0.54	2.5	22.8
8	7 +Therm. Mix. Valve	3.81	259	0.15	0.53	2.5	22.8

a. 380 Cycles/year

Table 7-11. Energy-use data, standard electric dryer (5.9 ft³), Procter & Gamble 1988–1989 data.

Level	Design Option	Energy Factor lb/kWh	Energy Use kWh/yr[a]
0	Baseline	2.60	967
1	Automatic Termination	2.95	851
2	1 + Insulation	3.01	834
3	2 + Recycle Exhaust	3.21	784
4	2 + Microwave	4.07	617
5	2 + Heat Pump	8.61	292

a. 359 cycles/year

clothes dryers, automatic termination of the drying cycle produces a 12% reduction in energy use relative to time termination. Increased insulation, from 1 to 2 inches, results in a 2% energy use reduction. Recycling exhaust heat can save 6% of energy consumption. The next two designs, microwave dryer and heat pump dryer, produce 26% and 65% energy savings respectively. Both of these designs are in the prototype stage (GMS Products 1989; Lewis 1987).

The heat pump dryer uses a room air-conditioner to dehumidify the air exiting from a conventional dryer. Much higher efficiency is attained, but at a significantly higher manufacturer cost. No heating element is needed. The exhaust air from the dryer passes through the evaporator coil of the dehumidifier where it cools down below the dewpoint, and sensible and latent heat are extracted. The heat is transferred to the condenser coil by the refrigerant and reabsorbed by the air in a closed cycle. A drain is required to remove the condensate. A prototype has been developed and successfully tested. It operates on 120 V and requires no vent because the air moves in a closed cycle. The prototype uses a disposable filter to reduce lint in the air system.

The microwave dryer employs two magnetrons (similar to those in microwave ovens) to evaporate water from clothes. Again, significant energy savings are achieved at significantly higher manufacturer cost. At the present time, there are some problems with electrical arcing for clothes with metal threads. Both of these technologies hold much promise, but additional research is needed to perfect them and to reduce their costs.

Other Products

The Department of Energy is analyzing eight products for potential standards updates: room air-conditioners, ranges and ovens, water heaters, direct heating equipment, pool heaters, mobile-home furnaces, fluorescent ballasts, and televisions. An Advanced Notice of Proposed Rulemaking was published by DOE on 28 September 1990. A Final Rule is expected to be published in 1992; the standards will take effect three years from that date.

New Technologies for the 1990s

Refrigerators and Freezers

A number of technologies or design options that would increase energy savings beyond the estimates discussed previously for refrigerators, freezers, and clothes washers were not incorporated into the final standards. Several of these new designs and their potential for increasing national energy savings will be described here. Section 4 will discuss policies for implementing such new designs.

For refrigerators and freezers, the use of vacuum insulation offers the greatest opportunity for increased energy savings. This design option was not included in the 1993 standard because it will require more time for research and development before it can be implemented on a commercial scale. One of the main issues is the reliability of mass-produced vacuum insulation panels. Additional laboratory and field testing is needed to determine if such panels can retain their integrity over 20 years or more. Another issue is whether the cost of manufacturing such panels can be reduced below present estimates of around $1 per square foot. The estimated cumulative (1993–2015) energy savings from implementation of vacuum insulation in refrigerators and freezers are very substantial, amounting to an additional 5.6 quads above the 5.2 quads achieved by the 1993 efficiency standards.

Clothes Washers and Dryers

A design option for both of these product types is to increase the spin speed during the spin dry cycle of a clothes washer. Typical U.S. washers utilize a spin dry speed of about 550 rpm. Many European washers have much higher spin speeds—well over 1,000 rpm for several models. Higher spin speed can remove substantially more water. Figure 7-5 shows the residual water in a 4.4-pound load of clothes as a function of spin speed (Asko Cylinda 1990). Increasing spin speed from about 550 rpm to 1,300 rpm decreased water retention from 50% to 25% of the starting water load. Thus, less total (clothes washer plus dryer) energy is consumed because

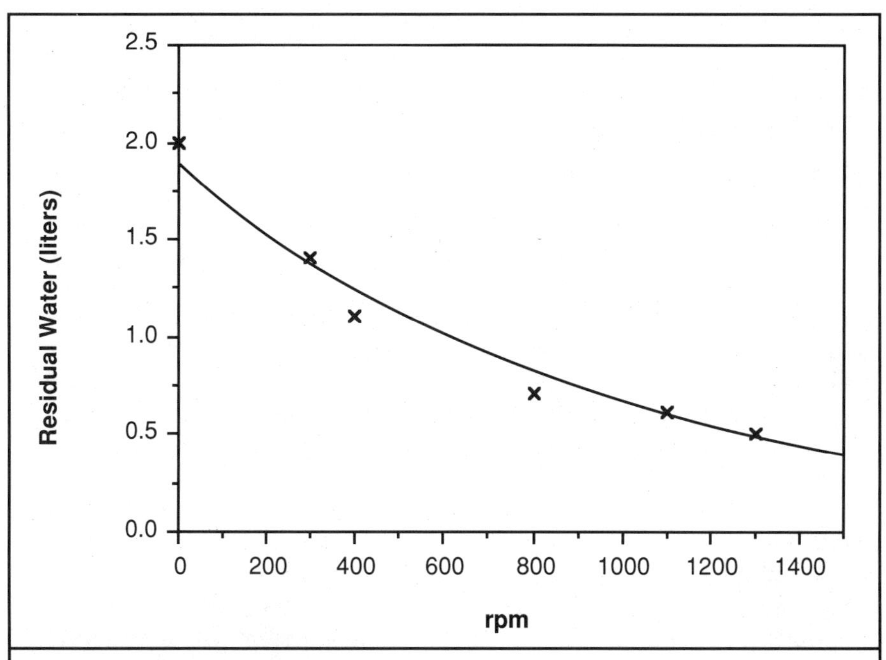

Figure 7-5. Residual water as a function of spin speed for a horizontal axis clothes washer.

much less energy is expended in mechanical water removal than in thermal water removal. Mechanical removal is about 70 times less energy intensive (135 lb water per kWh compared to 2 lb per kWh) than thermal removal in a dryer (DOE 1989b).

With the present DOE test procedure for clothes washers, there is no energy credit given for producing drier clothes at the end of the clothes washing process. Therefore, although this promising technology was analyzed during the rulemaking procedures, it was not included in the final standards. Changing the test procedure to allow for an energy-efficiency credit that depends upon the degree of dryness of the clothes washer load would motivate manufacturers to design for higher spin-dry speeds.

A possible change in the clothes washer test procedure is outlined in the Technical Support Document for Clothes Washers and Dryers (DOE 1989b).

Clothes washers with high spin speeds require additional attention during the design process in order to avoid load balancing problems and excessive noise. Another potential problem at high spin speed is that wrinkles may appear in the clothes. European manufacturers have developed programs that alter the spin speed during the spin cycle to avoid wrinkles.

Policy Mechanisms for Improving Energy Efficiency

There are many methods for fostering improved energy efficiency beyond the requirements of the present standards. Four are discussed below, and they are not mutually exclusive. One mechanism is already in place: the use of the federal standards update procedure to eliminate older, less efficient technologies as more efficient ones are proven to be commercially viable and economically justified. In this case, one relies on DOE to periodically (approximately every five years) increase the minimum energy efficiency of each product it regulates.

Several issues are important to consider if this path is used. One is the economic criteria used to establish standards. At present, trial standards with simple-payback periods of three years or less are usually incorporated into final standards. This policy assumes no adverse impact from the standards on consumers or manufacturers. Cumulative payback is calculated to include all design options that are part of the trial standard being evaluated. In these payback calculations, the DOE test procedure is used to calculate energy savings, but these procedures are unable to credit some new design options, such as higher spin speeds in clothes washers. It should also be noted that integrated appliances are not presently considered as products to be regulated by DOE. For example, the increasing popularity and higher water-heating efficiency of combined space and water heaters justify increased attention to such appliances. Test procedures should be updated to accommodate such new designs.

Another mechanism for promoting more efficient appliances is for utilities to offer rebates to consumers who purchase highly efficient appliances. Rebates are presently offered by many utilities in order to reduce electrical demand. A survey of 132 U.S. utilities showed that 60% offered rebates for heat pumps, 40% for central air-conditioners and water heaters, and 27% for refrigerators (EPRI 1987). The rebate amounts were \$110–\$300 for heat pumps and \$30–\$50 for refrigerators. Efficiency standards established for many products may reduce the range of efficiencies manufacturers produce, making rebate guidelines more difficult to determine. The range of efficiencies for refrigerators listed in the 1990 AHAM Directory of Refrigerators and Freezers has narrowed considerably from 1989. However, if utilities continue to offer rebates for purchasing ever more efficient appliances, manufacturers may be motivated to produce such products.

Recently in California, several utility companies have offered rebates of $50 and $100 to consumers purchasing refrigerators that are at least 10% and 15% respectively more efficient than required by the 1990 standards. Manufacturers have responded positively to such programs by producing greater quantities of their more efficient models for sale in California. California utilities can now earn a profit based on the amount of energy saved by various conservation programs. Another example of such a program is the Blue Clue program sponsored by Bonneville Power Administration (Anderson 1990), which uses labels to tell consumers which refrigerators and freezers are among the top 15% of all models in energy efficiency. Several other utilities have adapted this approach. Information programs of various kinds can be very helpful in increasing consumer awareness of the energy and environmental impact of their appliance purchase decisions.

A third mechanism is for the federal government to offer manufacturers incentives to produce, and consumers incentives to use, more efficient appliances. For example, manufacturers could be offered a tax credit based on a weighted average efficiency of products manufactured. If that efficiency is high enough (relative to the existing DOE efficiency standard), then corporate taxes could be reduced on a sliding scale dependent on the difference between these two efficiency values. Congress would have to approve such a policy, which would likely be quite controversial. Also, consumers could be given a tax credit or a direct rebate to induce them to purchase more efficient models. Similar to utility rebates, this practice would serve to increase demand for more efficient products.

A fourth possible approach is for the federal government to assist industry with research on key technologies that are needed for significant efficiency improvements. This could be done by increased spending in national laboratories and universities or through joint research efforts by industry and government. One such effort is now under way to address CFC issues in the refrigerator industry; for this purpose, a unique consortium (Appliance Industry-Government CFC Replacement Consortium) has been created as an AHAM subsidiary and is made up of refrigerator and freezer manufacturers, the DOE, the Environmental Protection Agency, compressor manufacturers, and chemical suppliers. A number of technologies and issues would benefit from more research and development efforts. For example, research involving different detergents and soils could clarify the debate concerning the proper temperature for dishwashing. As mentioned earlier, higher spin speeds for clothes washers could save energy in the clothes-drying process. DOE could perform more research on altering the present test procedure for clothes washers and dryers with the goal of promoting clothes washers that remove moisture more efficiently. Evacuated panels for refrigerators, microwave dryers, heat pump dryers, induction cooktops, and horizontal axis clothes washers are examples of technologies that require more research and development.

Conclusions

Several recommendations can be distilled from the above discussion:

1. DOE should receive increased funding for research on appliance efficiency.
2. Additional industry/government research efforts are desirable.
3. Utility rebate programs should be continued.

Congress should direct DOE to conduct more research in the area of appliance energy efficiency. Increased budgets to carry out that work will be necessary. Laboratory and field use data are needed for several areas of technology as described above. This research could be carried out at the National Institute for Standards and Technology (NIST), national laboratories, universities, utilities, or manufacturers' facilities. Most productive are combined industry and government efforts that force the regulators and regulated to work together for a common goal. Field data are needed to adjust DOE test data so that the latter more closely represent actual energy use in residences. This step would improve the simple-payback calculation discussed above. Utility rebates to consumers for the purchase of the most efficient appliances should be continued. These rebate programs provide incentives to manufacturers and motivate consumers to purchase more efficient products.

In conclusion, the key to continued improvement in appliance energy efficiency is better cooperation between DOE (and its contractors) and appliance manufacturers. Manufacturers possess the practical experience required to add new features to their products, while DOE is required to periodically revise the existing efficiency standards, if technologically feasible and economically justified designs can be identified. Any revisions must be achievable without not reducing the utility of the product or manufacturers' profits. Recently, manufacturers have begun to invest the money and time needed to bring new energy-efficient technologies to the marketplace. It is hoped that the importance of energy efficiency as an inhibitor of global climate change and air pollution will prompt a greater interest in efficiency on the part of appliance manufacturers and their suppliers. Such efforts will also make the U.S. more competitive in the international marketplace.

Acknowledgements

This work was supported by the Assistant Secretary for Conservation and Renewable Energy, Office of Building Technologies, Office of Codes and Standards, of the U.S. Department of Energy under Contract No. DE-AC0376F00098. Opinions expressed in this chapter are solely those of the authors and do not necessarily represent those of DOE or Lawrence Berkeley Laboratory.

Appendix: Energy-Efficiency Standards for Other Products

Room Air-Conditioners (Effective 1 January 1990)

Size (kBtu/hr)	EER*	EER**
< 6	8.0	8.0
6– 7.999	8.5	8.5
8–13.999	9.0	8.5
14–19.999	8.8	8.5
≥20	8.2	8.2

* Product class: without reverse cycle and with louvered sides
**Product class: without reverse cycle and without louvered sides

Note: EER, the energy efficiency ratio, equals capacity in Btu/hr divided by electrical power input in W.

Central Air-Conditioners and Heat Pumps

Product class	SEER	HSPF
Split system	10.0	6.8 (Effective 1 January 1992)
Single package	9.7	6.6 (Effective 1 January 1993)

Note: SEER is the seasonal energy efficiency ratio. HSPF, the heating seasonal performance factor, is the heating capacity in Btu/hr divided by electrical power input in W.

Water Heaters (Effective 1 January 1990)

Product class	Standard
Gas-fired	0.62 − (0.0019 × tank vol. in ft^3)
Oil-fired	0.59 − (0.0019 × tank vol. in ft^3)
Electric	0.95 − (0.00132 × tank vol. in ft^3)

Furnaces

For central furnaces, annual fuel utilization efficiency (AFUE) must be at least 78% by 1 January 1992.

For mobile-home furnaces, AFUE must be at least 75% by 1 January 1992.

Ranges and Ovens

Gas kitchen ranges and ovens having an electrical supply cord and manufactured after 1 January 1990 shall not be equipped with a constantly burning pilot light.

References

AEG. 1989. Product literature on alternate spray arms in Oeko-Favorit dishwasher. Nurnberg, Germany.

AHAM. See Association of Home Appliance Manufacturers.

Anderson, K. 1990. "Appliance Efficiency Programs: Beyond Rebates." *Home Energy* 7 (March 1990): 13–17.

Arthur D. Little, Inc. 1982. *Refrigerator and Freezer Computer Model User's Guide*. Cambridge, Mass.: Arthur D. Little, Inc.

Asko Cylinda. 1990. Product literature from 1990 Domotechnica Exposition in Cologne, West Germany. Asko Cylinda Co.

Association of Home Appliance Manufacturers. 1989. *1989 Directory of Certified Refrigerators and Freezers*. Chicago: Association of Home Appliance Manufacturers.

——. 1990. *1990 Directory of Certified Refrigerators and Freezers*. Chicago: Association of Home Appliance Manufacturers.

DOE. See U.S. Department of Energy.

EIA. See Energy Information Administration.

Electric Power Research Institute. 1987. *A Compendium of Utility Sponsored Energy Efficiency Rebate Programs*. EM-5579. Palo Alto, Calif.: Electric Power Research Institute.

Energy Information Administration. 1989. *Annual Energy Outlook, 1989*. DOE/EIA-0383(89). Washington, D.C.: U.S. Department of Energy.

EPRI. See Electric Power Research Institute.

GEA. 1989. "Response to DOE's Notice of Proposed Rulemaking Regarding Refrigerators." 31 January.

GMS Products. 1989. "Comments on Notice of Proposed Rulemaking." 10 October.

Lebot, B., I. Turiel, and G. Rosenquist. "Horizontal Axis Domestic Clothes Washers: An Alternative Technology That Can Reduce Residential Energy and Water Use." In *Proceedings of the 1990 ACEEE Summer Study on Energy Efficiency in Buildings.* Vol. 1. Washington, D.C.: American Council for an Energy-Efficient Economy.

Lewis, D. 1987. "The Nyle Dehumidification Residential Clothes Dryer." Bangor, Maine: Nyle Corporation.

McMahon, J. 1987. "LBL Residential Energy Model: An Improved Policy Analysis Tool." *Energy Systems and Policy* 10 (1): 41–71.

Potter, T. 1989. Solar Energy Research Institute, Golden, Colorado. Letter to I. Turiel, 11 December.

Procter & Gamble. 1989. "Comments on Notice of Proposed Rulemaking." 5 October.

Quantum Optics. 1989. "Silica Aerogel Refrigeration Insulation." Presentation by Quantum Optics, Inc., at U.S. Department of Energy public hearing on 12 January.

U.S. Department of Energy. 1988. Technical Support Document: "Refrigerators, Refrigerator-Freezers, and Freezers; Small Gas Furnaces; and Television Sets." DOE/CE-0239, November.

——. 1989a. *Federal Register* 54 (17 November): Vol. 54, 47, 916–47, 945.

——. 1989b. *Technical Support Document: Dishwashers, Clothes Washers, and Clothes Dryers.* DOE/CE-0267. Washington, D.C.: U.S. Department of Energy.

——. 1989c. Technical Support Document: "Refrigerators and Furnaces." DOE/CE-0277, November.

U.S. Department of Energy/Energy Information Administration. 1989. *Household Energy Consumption and Expenditures 1987.* DOE/EIA-0321. Washington, D.C.

White Consolidated. 1989. Data submitted to the U.S. Federal Trade Commission on models LT250L, LT700L and LT800L. 13 February.

Chapter 8

Looking Beyond Aggregate Household Energy Demand: What Really Happened to Conservation

Andrea Ketoff and **Lee Schipper,** *Lawrence Berkeley Laboratory*

Changes in energy demand in the residential sector constitute an important part of the overall evolution of energy consumption in the countries of the Organization for Economic Cooperation and Development (OECD) since the oil shock of 1973. Abrupt price increases and the implementation of a multitude of energy-saving policy measures changed households' energy-use patterns. Conversely, in recent years, the fall of oil prices and the expiration of many conservation policies have stimulated an increase in household energy demand.

With concerns mounting about the contribution of the United States and other industrialized countries to the global emission of greenhouse gases, a thorough understanding of the dynamics of changes in fossil-fuel demand is imperative if we are to establish reduction potentials and targets. In the United States, Japan, and Europe, residential energy demand constitutes 16-32% of overall primary energy consumption.[1] The United States alone, with 40% of the population of the OECD, accounts for almost 60% of the primary energy consumption under consideration. By comparison, energy

---

1. The scope of this study is limited to seven European countries (Denmark, France, West Germany, Italy, Norway, Sweden, and the United Kingdom). We refer to the aggregate of these countries as Europe-7, or Europe. The United Kingdom includes England, Scotland, Wales, and Northern Ireland.

use in the Japanese residential sector is only 13% of that in the United States. The per capita levels shown in Figure 8-1 indicate that in 1987, the average American consumed close to four times more primary energy than the average Japanese for home comfort, and almost twice what the average European uses. This difference has been closing in the last 15 years but is still large.

Comparing changes in energy use is important to evaluating past performance and estimating future trends for each of the major countries. The evolution of aggregate energy consumption—whether measured in primary, delivered, or useful terms (see footnote 3 for definitions of these terms)—hides the complex changes in components of consumption. After growing steadily in the 1960s and early 1970s, household energy demand in all studied countries except Japan has been relatively flat from 1972 to 1987 (Figure 8-2). Yet this stability resulted from enormous reductions in energy use per unit of activity for space heating and certain appliances, balanced by significant increases in energy use for the same applications, caused

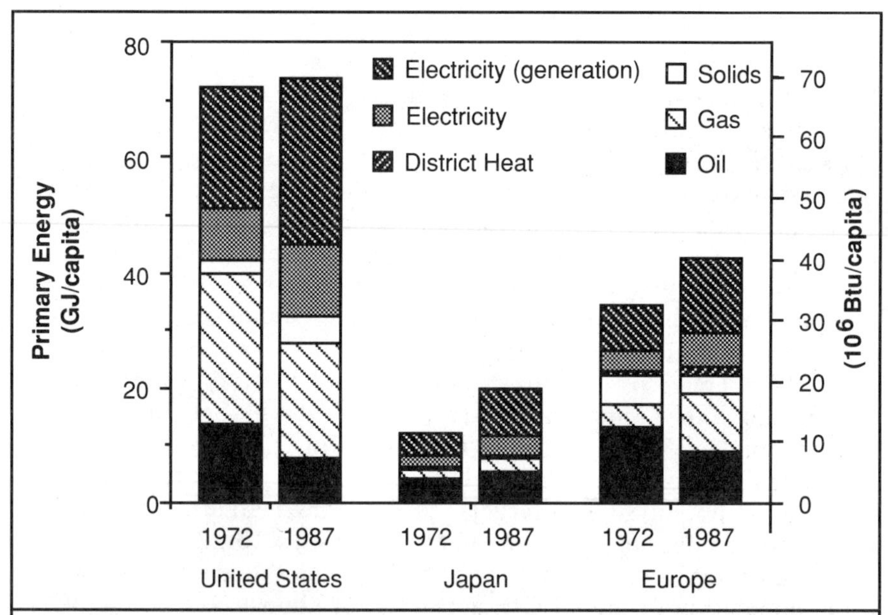

Figure 8-1: Primary residential energy demand: United States, Japan, and Europe.

Note: Europe includes West Germany, France, Italy, the United Kingdom, Denmark, Sweden, and Norway.

1 GJ = 0.958 × 10^6 Btu

by increases in living standards and acquisition of more heating and appliance equipment.

The goal of this study, therefore, is to quantify what really happened to household energy use since the early 1970s. This quantification will allow us to estimate the impacts of more efficient energy use—energy conservation—on total residential energy use in each studied country. We will estimate how much of the change in household energy use arose from technical changes that cannot be reversed easily in an era of low energy prices. We will be able to use these findings to judge which forces might decrease or increase household energy use in the future. Finally, we will be able to place changes in U.S. household energy use in an international context,

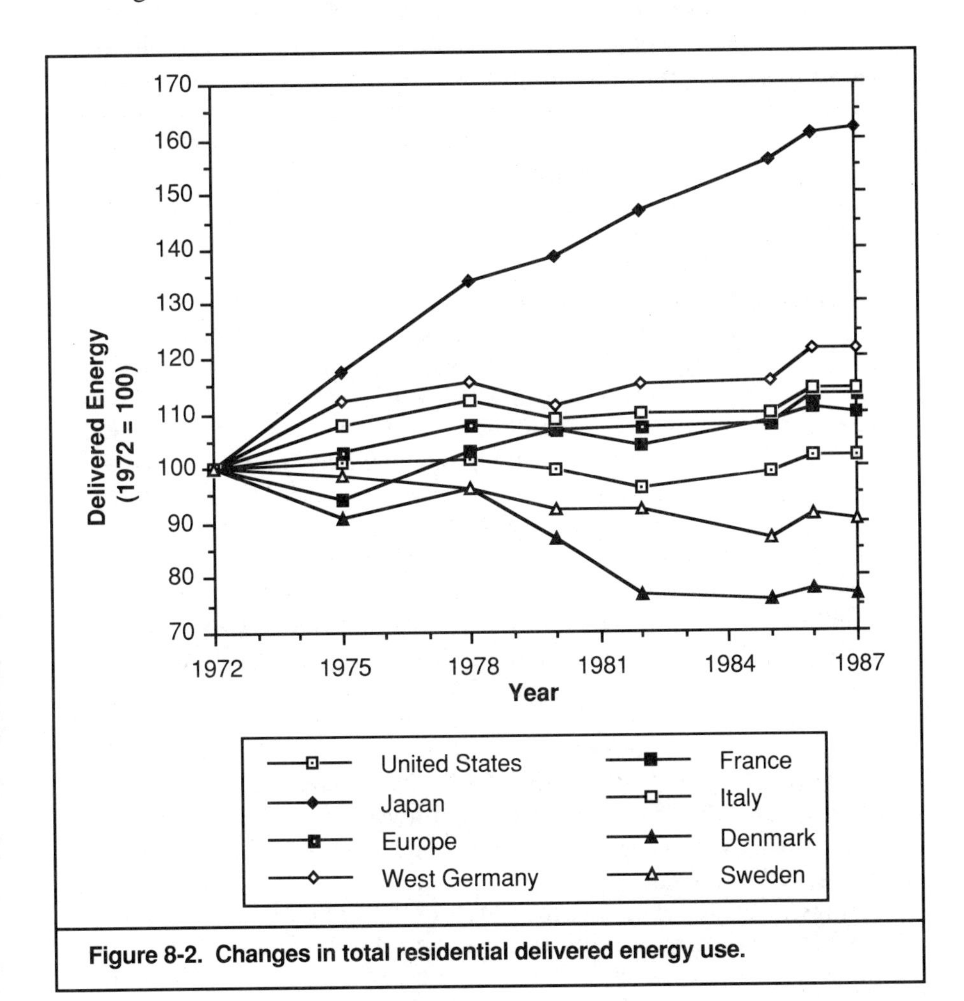

Figure 8-2. Changes in total residential delivered energy use.

through comparison of how each component of use in the United States evolved vis-à-vis similar changes in other countries.

We define conservation as a purposeful reduction in energy use per unit of activity (Schipper and Ketoff 1985). This reduction can be achieved by increasing the efficiency of energy-using devices, such as furnaces, refrigerators, or building envelopes. Additionally, conservation is achieved by reducing consumer demand for specific services, such as space heating or hot water. Conservation occurs in response to higher prices, information, programs, or other forces.

Background

Energy conservation is a set of responses to price and policy signals, responses that typically result in reduced energy use in the home. These responses must be monitored over time for several reasons. First, consumer reactions occur after a time lag, and the nature of the response and its direction can change over time. Second, price and policy signals have been unstable over time. Finally, the impacts of uncertainties in data and of long-term changes not closely related to energy conservation can only be separated from the impact of conservation if our observation period is long enough to allow the effects of conservation to accumulate.

The analysis of responses to price and policy over time permits us to distinguish between permanent and reversible changes in energy use. Permanent changes result from investments in increased efficiency, fuel switching, and the introduction of new, more efficient systems. These changes can be undone only over long periods of time, after new equipment is worn out and replaced by less efficient equipment. Reversible changes such as lower thermostat settings are driven by consumer behavior that can be introduced rapidly with almost no investment cost; these changes can be undone overnight. The analysis of any possible upturn in energy use will help us to quantify the short-term reversibility of consumers' efforts to save energy.

In previous work, we evaluated the permanent and reversible components of oil savings in the residential sector of major OECD countries (Schipper and Ketoff 1985). This study suggested that by 1983 seven OECD countries had achieved overall oil savings of around 40% with respect to the level expected (assuming no changes in the 1972 equipment and energy intensity). The residential oil savings in these seven countries amounted to nearly 1 million barrels of oil per day, making the residential contribution one of the most significant in the overall reduction in OECD oil demand

(Schipper 1987). Also, we estimated that 45% of these savings were permanent, while the rest could reverse if oil prices and conservation programs were to relax the pressure on consumers. This reversal is currently under way. Important breaks in consumption patterns occurred in the period 1983–1987, when the international oil price weakened and then fell, and conservation policies relaxed or ended. The oil price crash of 1986 accentuated this drop in prices. In order to analyze the nature of this reversal, this chapter focuses not only on the differences in end points of each country's consumption, but also on the differences in paths that consumption in each country has taken.[2]

The historical analysis is applicable to the future if we first identify and understand the components of changes in household energy use, and then try to assign causes to these changes. These components include

- Structural changes (changing size of dwellings and varying types of housing)
- Changes in equipment characteristics (types of furnaces, relative saturations of central heating systems and stove heating, and saturations and energy-related characteristics of appliances)
- Changes in the characteristics of building shells
- Changes in household behavior
- Effects of fuel-switching

The causes of these changes could include higher prices, changing incomes, conservation programs, building and appliance standards, and new technologies. Before these causes can be discussed, however, the magnitude and direction of the changes must be identified; this latter task is one of the principal goals of this chapter.

The aggregate picture of household demand does not permit this depth of analysis. Aggregate demand (or aggregate change) obscures the role of the various components of change, many of which have offsetting impacts on total household demand. We cannot, for example, distinguish the impact of changes that are income-driven and often the result of long-term lifestyle changes (larger homes) or demographic trends (changes in family size) from those that are directly related to energy prices or various conservation policies and that are thus possibly reversible. Without such a distinction, we cannot measure the impact of energy conservation. Lacking a clear picture of historical changes and their importance

---

2. For comparisons between these countries, as well as for analysis of individual countries, see our earlier work (Schipper and Ketoff 1983; Schipper et al. 1985; Ketoff et al. 1987).

today, we would be unable to bracket likely or plausible changes from energy conservation in the future. But with such a disaggregation, we can determine the separate influences of each important component of demand, and even estimate what the level of energy demand would have been if the major components had changed one at a time. Our methodology is essentially "bottom up," quantifying all the components of household energy use that together create total demand by fuel and end use.[3]

To identify the components and causes of changes in household energy use, we disaggregate residential demand by different end uses, by fuels, and, when data permit, by types of systems in which energy is used (central and noncentral heating, single-family or multifamily dwellings, newer [post-1973] or older homes). We focus on energy use for space heating, the most important end use in the countries examined, and electric appliances, which grew significantly in almost all countries (Figure 8-3, page 236). The share of household appliance electricity for residential air-conditioning is important in the United States (18%) and Japan (11%), but negligible in other countries. Because cooking accounts for less than 10% of energy use in all countries, and because most of the figures available for water heating are rough estimates extracted from consumption in combined space and water heating systems, we attribute most of the variation in household energy consumption to space heating.

Our analysis covers the period 1972–1987. Because changes in energy use and its components have not been monotonic or continuous over the period observed, we present data from intermediate years to show important variations in trends, although this presentation is limited to a certain extent by lack of consistent data in some countries.[4]

---

3. We measure energy in three ways. *Delivered (or secondary) energy* corresponds to consumption at the building boundary, excluding the conversion of primary energy sources into electricity or district heat. Delivered energy is thus the consumer's billed energy, plus energy derived from gathered wood. We also measure *useful energy* to compare household requirements for space heating and water heating when both electricity or district heat—for which there are no combustion losses—and fossil fuels with combustion losses are used. Useful energy comes closest to measuring the levels of energy services delivered by different fuels; the following efficiencies are considered for calculating useful energy from delivered energy: 66% for oil and gas, 55% for solids, and 100% for electricity and district heating. Finally, *primary energy* accounts for the conversion of primary energy sources into those actually used by the household sector. In our work, this means that losses in producing district heat and electricity are included; to the extent that these sources are produced with fossil fuels, primary energy use is a good index of the emission of CO_2.

4. We rely wholly on energy consumption data from individual countries. The

Space Heating

Space heating is the largest component of household energy end use in every country except Japan. Yet, as shown in Figure 8-3, the relative importance of space heating declined in all countries between 1972 and 1987. This important change was caused both by the sustained growth of other end uses, particularly electricity use for appliances, and by the decline in energy use per dwelling for space heating (Figure 8-4, page 237). Changes in the characteristics of homes, in the fuels used for heating, and in the intensity of space

---

published energy consumption data from the International Energy Agency (IEA) do not accurately represent the residential sectors of most countries, as we found by comparing IEA data (which are not referenced or annotated) to individual country data (IEA 1987a; IEA 1989a). The data sources deserve certain caveats. In most countries, "data" consist of national energy balances estimating the use of each fuel in the residential sector, separate surveys of equipment ownership, and time series of household and housing characteristics. But these balances are usually based on reports from energy suppliers, reports that are not always reliable. Only a small number of countries have conducted regular household fuel use and equipment surveys covering all fuels: the United States since 1978 (EIA RECS) and France (a private, unpublished survey carried out yearly since 1975 for government authorities). In a few other countries (Germany, the United Kingdom, Sweden, Japan, Denmark) there are estimates of total residential energy use published by recognized energy authorities but no regular surveys to support these estimates. Authorities in Sweden do survey the use of heating fuels, but until recently ignored water heating and electricity for nonheating purposes.

Oil use poses special problems. Household oil consumption is only measured regularly in the United States, France, Sweden, and Germany. In countries where oil plays an important role in the household sector and no measurements are taken, the residential sectors still are absent from official yearly energy balances. Utility gas and electricity is easier to classify, but meters are rarely read more than once a year, except in the United States, Japan, and Canada. Thus, for many countries national data on the consumption of household energy are inconsistent: reliable for some fuels, unreliable for others. By the late 1980s, however, energy consumption data were being published regularly by a recognized authority in every country except Italy. The Italian national utility—ENEL—surveys household electricity use with regularity, but little is known about other fuels. We use a variety of country sources to estimate the share of each fuel for each main end use. We also make use of surveys of household and housing characteristics and equipment ownership from each country, surveys not found in any international sources (Schipper et al. 1985). In some contexts, we aggregate the seven European countries to make simplified comparisons with the United States or Japan. Only by using disaggregated data are such aggregations possible. The data developed within this 12-year project at the Lawrence Berkeley Laboratory have become a widely accepted reference source for international institutions, governments, utilities, and oil companies worldwide. Among the sponsors of this analysis of residential energy use in OECD countries are the International Energy Agency (IEA), the Royal Norwegian Ministry of Oil and Energy, the Swedish Council for Building Research, the Italian National Committee for Reasearch and Development of Nuclear and Alternative Energy (ENEA), and the Oslo Electric Utility (Oslo Lysvaerker). These data have also been used since 1984 by Shell International Petroleum Co., British Petroleum, Exxon Corp., and Agip SpA.

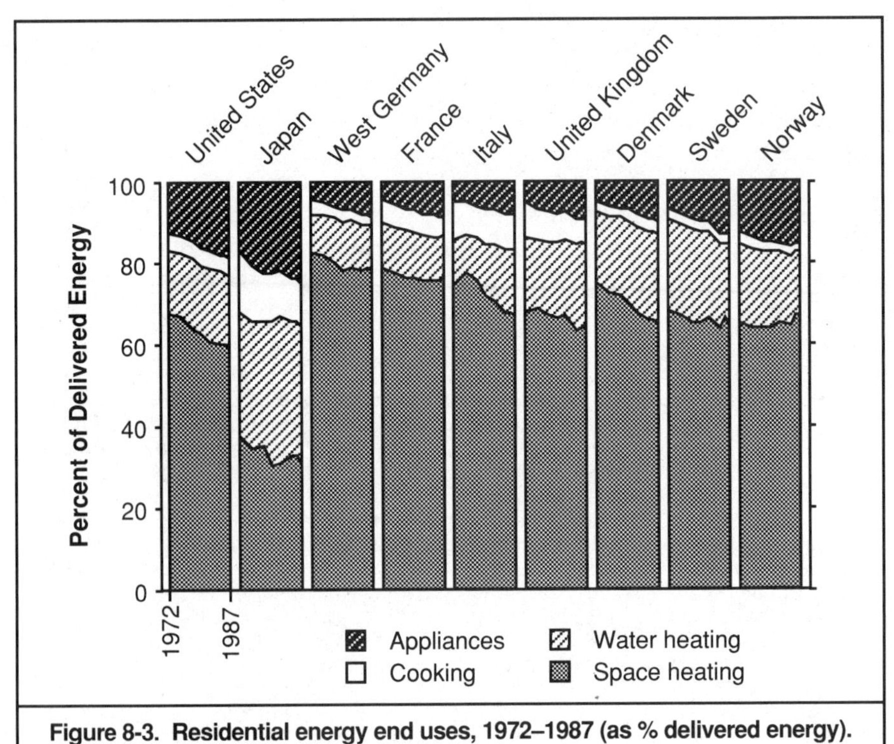

Figure 8-3. Residential energy end uses, 1972–1987 (as % delivered energy).

heating caused the overall changes. When we examine these changes separately, we find that the impact of reduced heating intensities in most countries was considerably greater than indicated by the heating per household figures.

Structural Changes in Space Heating Demand

The characteristics of dwellings, households, and heating systems, as well as the severity of the climate (measured in degree-days), define the structure of residential heating demand. The number of people in a home, the size of the dwelling, and the heating system characteristics influence space heating demand. Significant changes in these structural components during the period 1972–1987 have critically affected space heating demand. We thus need to consider these changes in detail.

From 1972 to 1987, population growth ranged from −1.0% (in West Germany) to +16.3% (in the United States), while household size decreased by 12–14% in the countries studied. Together, these changes led to an important increase in the number of

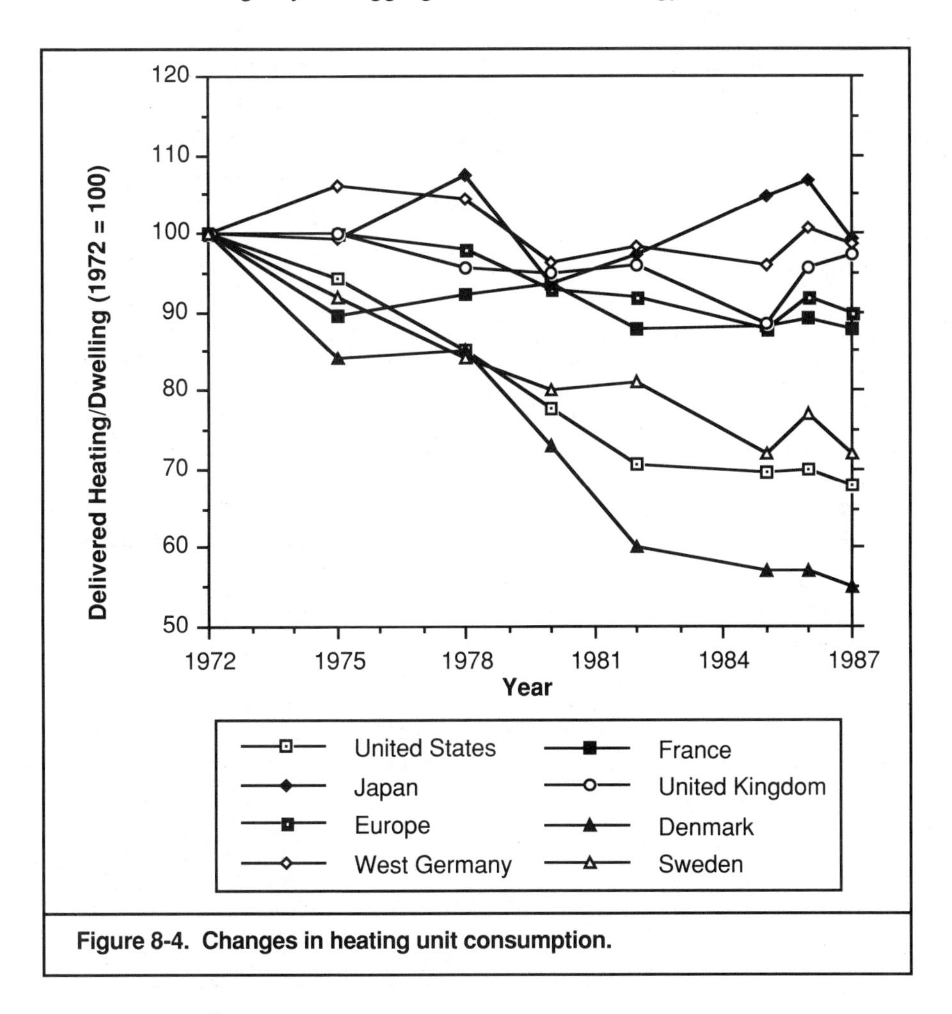

Figure 8-4. Changes in heating unit consumption.

dwellings and, therefore, in area heated. Table 8-1 (next page) shows the increase in population and housing stock (both much higher in the United States and Japan than in Europe) and the decrease in average household size for the same countries. These demographic changes also increased equipment ownership per capita, leading to higher per capita energy consumption. In cross-sectional comparisons of household energy use (normalized by household size), per capita energy use is largest in the smallest households. In fact, although energy consumption per dwelling may drop, reduced household size in the United States (and all OECD countries) led to increased residential energy use.

The combined effect of demographic changes since 1972 has

Table 8-1. Changes in socio-demographics (1972 to 1987).

	Population Growth[a]	Household Size 1972	Household Size 1987	Housing Stock Growth[a]
United States	1.0%	3.1	2.7	1.9%
Japan	0.8%	3.8	3.3	1.8%
Europe[b]	0.2%	3.1	2.6	1.1%

a. Average annual growth rate
b. Seven European countries included in this study

increased household energy use by more than 30% in both the United States and Japan, and by around 20% in European countries. Among these European countries, the impact of demographic changes was largest in Italy (26%) and smallest in the United Kingdom (13%).

Dwelling size increased throughout the study period. Together with the decrease in household size, the available space per person grew significantly in each country. Figure 8-5 shows the growth in dwelling area per capita (we presume the added space was heated) that occurred as a result of both these changes. This growth varied from between 1.4% per year in Denmark to 2.1% per year in Japan. This growth increased space heating demand per capita. The hypothetical impact of growth in dwelling size alone on 1972 space heating demand was more important in Japan and in those European countries where average dwelling area was smaller in the early 1970s (Italy, France, Sweden) than in the United States, where dwelling size did not increase as much. However, the growth in dwelling area contributed less to the change in demand than did either the reduction in household size or that in growth in homes per capita.

Indoor thermal comfort is best represented by a complex function of indoor temperature, humidity, air movement, and other parameters. Detailed surveys estimate the share of a home's volume actually heated, the indoor temperatures aimed for (or achieved), and the hours of heating. But such information is rarely available on a national scale and covering an extensive period. We can approximate the national comfort level by looking for measures of the amount of heating that can take place in a home. Ownership of central heating systems comes closest to such measure. Central heating permits all rooms to be heated to the desired temperature. Homes with central heating systems, in which heat is delivered to

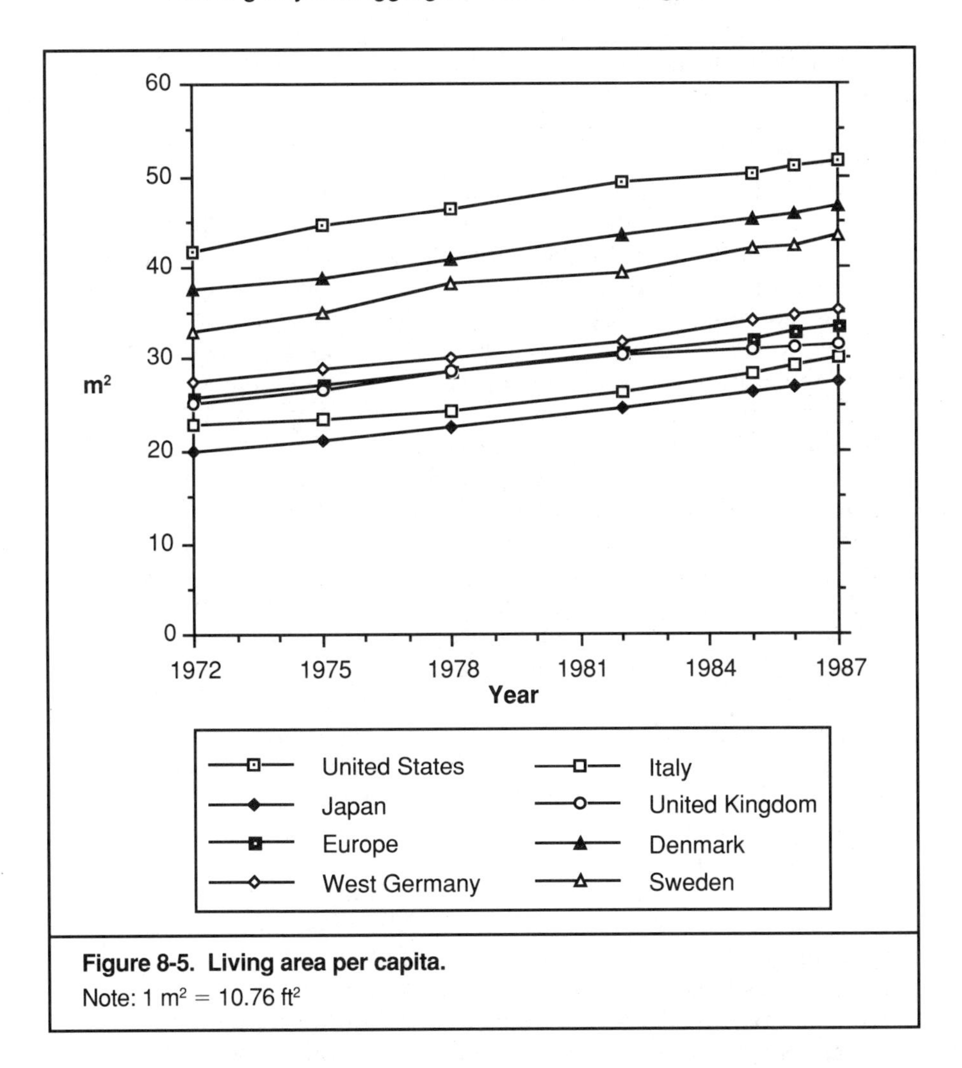

Figure 8-5. Living area per capita.
Note: 1 m² = 10.76 ft²

every room from a boiler or furnace (or from electric heaters in every room), typically consume two to three times more fuel than their room-heated counterparts, where small gas, wood, coal, or kerosene stoves are used (Schipper et al. 1985; IEA 1987b).

To calculate the impacts of changes in central heating saturation on energy use, we assume that centrally heated homes use two times more fuel than homes heated by stoves. Data from individual gas and oil companies in Europe and the United States bear out this approximation. We then estimate the impact of greater central heating penetration by calculating what space heating demand would be if only the saturation of central heating systems had changed, with housing stock and intensities of central and noncentral systems remaining at 1972 levels (Ketoff and Schipper 1990).[5]

The impact of changes in central heating saturation was minimal in the United States, Sweden, and Denmark, since these countries already had high central heating penetration in the early 1970s (see Figure 8-6). On the other hand, changes in central heating saturation had important impacts on demand in the four large European countries (Germany, France, Italy, and the United Kingdom) where saturation increased from around 40% of the housing stock in the early 1970s to nearly 70% by 1987. Overall, if only central heating saturation had changed since 1972, aggregate space heating consumption of the seven European nations in this study would have increased around 15% by 1987.

In Japan, where central heating is present in only 6% of dwellings, the central-heat calculation cannot be applied. The same is true for Norway, which has a low penetration of central heating. Still, 65% of all homes have electric radiators in all major rooms, and another 15% have either radiators or wood or kerosene stoves in every room. In terms of comfort, we can consider these systems as central heating because they provide the same heating comfort as do true central systems. In Japan, room systems fueled principally with kerosene (or occasionally gas) are used when rooms are occupied, but turned off otherwise. Electricity is used as a complement to kerosene. Traditionally, every home has one or more foot-warmers. Now, nearly 50% of all homes have acquired small heat pumps (which function as air-conditioners in the summer) to provide heat during the milder winter months. More recently, electric heat pumps have added comfort in the spring and fall, when outdoor temperatures are mild. However, indoor temperatures in Japan during the heating season are still well below those in Europe or North America (Energy Conservation Center 1989). While heating comfort in Japan increased markedly in the 1970s and 1980s without a significant penetration of central heating systems, households have not reached the level of thermal comfort common in the other countries in this study.

The impact on space heating demand of changes in the structural components analyzed above (population growth and household size, dwelling area, and central heating saturation) is compared

5. We estimate the energy intensity of central and noncentral heating systems for 1972 (assuming a standard difference of a factor of two between centrally heated homes and homes heated by stoves), and then calculate hypothetical total space heating energy use of the 1972 housing stock at 1987 levels of central heating saturation. For detailed formulas, see Ketoff and Schipper 1990.

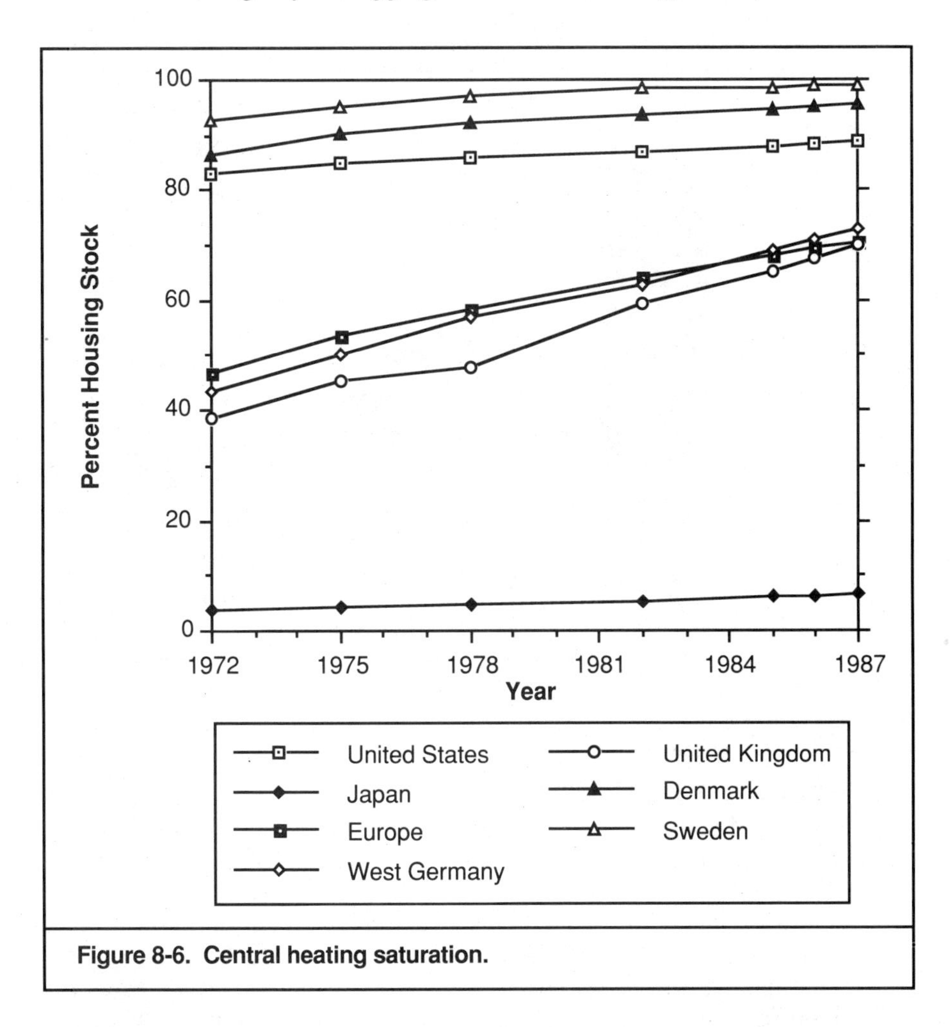

Figure 8-6. Central heating saturation.

in Figure 8-7 (next page) for the United States, Japan, and Europe. Overall, the effect of demographic changes constituted by far the strongest upward pressure on space heating demand, but increases in dwelling size and in central heating saturation (in the case of Europe) were also important. Clearly the overall impact of structural change was significant; in principle, the combined effect of these changes could obscure or even offset all conservation efforts.

All the structural changes examined here can be considered permanent. Reductions in dwelling size or increases in household size do not occur in the short term. Similarly, households are unlikely to abandon central heating equipment once it is installed, but might cut back on energy use if fuel prices increased rapidly,

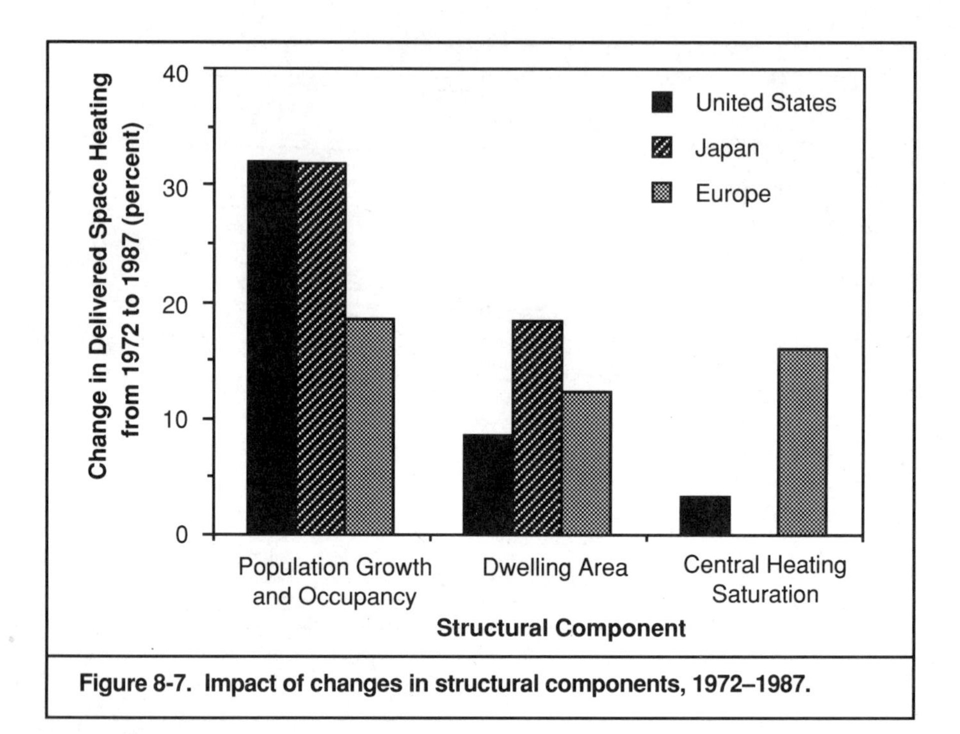

Figure 8-7. Impact of changes in structural components, 1972–1987.

or if incomes suddenly fell (as happened in Denmark between 1981 and 1983). Such cutbacks appear as changes in unit consumption or intensity. At the same time, many of the structural changes will approach saturation or asymptotes in size (a "largest" home size, a "smallest" household size) and in equipment ownership and even use. This implies that many of the components of household energy use that led to strong growth in demand may not yield strong growth in the future. As a result, conservation efforts may be more visible in the future than they were in the past.

The Effect of Fuel-Switching

Fuel choice influences total space heating energy demand, because different fuels can be converted to energy services with different efficiencies. Although we do not count fuel-switching per se as energy conservation, it is important to estimate the impact of switching on total use. Therefore, we must review changes in actual fuel choice to see whether they have influenced energy use over the observation period.

Figure 8-8 (page 244) shows the dramatic changes in fuel choices for space heating that occurred in 1972-1987. The share of dwellings heated with oil and solid fuels was reduced in all countries. This substitution occurred through three processes:

1. Most households converted over time from dirtier, bulkier fuels (principally coal, but also wood or oil) to cleaner, more compact ones (often oil or gas, but also electricity or district heat). Usually, the switch away from solid fuels (or oil products like kerosene) occurred together with the switch from stove heating to central heating. This was the case for Italy, Germany, and the United Kingdom, where use of solids was reduced markedly. By contrast, solids—particularly coal—had lost their importance in the United States by 1960 and in Scandinavia before 1970. In all these countries, households with solid fuel-based central heating switched to oil or gas or both, often with the same basic heating system remaining in place.
2. Gas, electricity, and, in some countries, district heating have been the dominant fuels used for heating in new construction after 1973, although the share of oil persisted above 25% in Germany, France, and Denmark until the early 1980s.
3. Households switched fuels in existing homes to reduce heating costs, as a result of changes in relative prices (or policies that juggled prices or subsidized switching). In both Sweden and Norway, electricity became popular when oil prices increased in 1979. Natural gas was substituted for oil in Italy as a result of incentives from gas utilities; also important in Italy was the economic convenience of dwelling-size gas-fueled furnaces that allowed independent regulation and management, in contrast to building-size systems that allow little flexibility. The almost universal presence of central heating in Sweden facilitated the acceleration of switching from oil to district heating. In the United States, high oil prices brought rapid switching to natural gas, accelerating an existing trend, or to wood, reversing instead the previous trend. In all these cases, the substitute was either very cheap or readily available with little inconvenience to building occupants. In some cases, fuel-switching occurred in a reversible manner: homes maintained the capability to use oil or wood. In Germany, on the other hand, switching in dwellings already equipped with central heating was insignificant due to the small price differential between oil and natural gas and to the relative newness of oil-fired equipment.

As Figure 8-8 shows, the final results of fuel-switching have been dramatic in all countries. Oil (and LPG) lost a considerable fraction of its share of heated homes. District heating and electricity

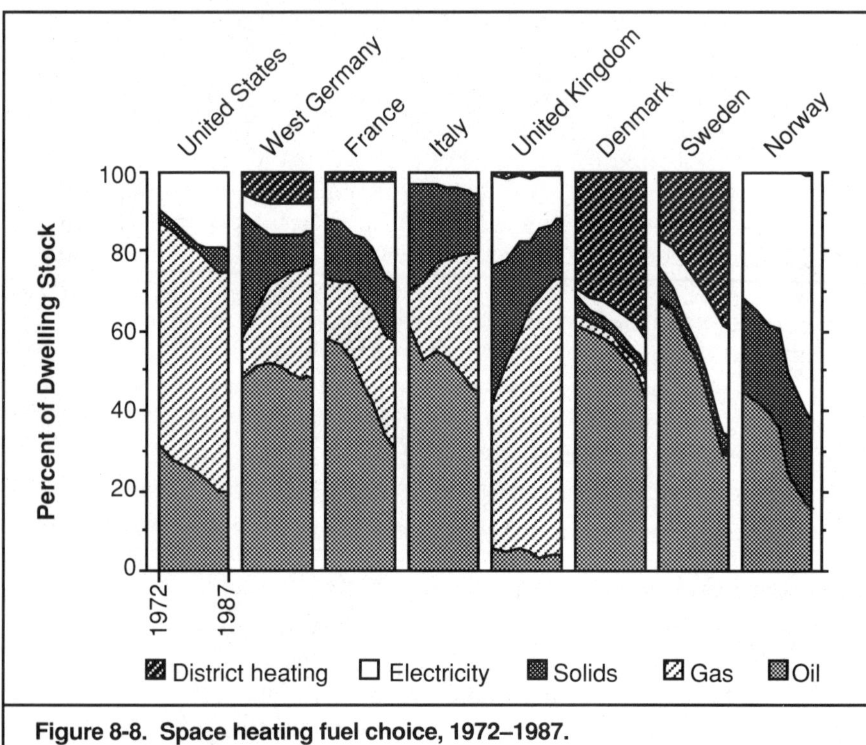

Figure 8-8. Space heating fuel choice, 1972–1987.

were the most important substitutes in the Scandinavian countries, while gas, electricity, and wood predominated elsewhere. (Figure 8-8 does not show the remarkable move away from oil toward solids, gas, and renewables as primary sources for district heating systems in Sweden and Denmark.)

Natural gas more than doubled its share of homes heated in Germany, Italy, and the United Kingdom, substituting for coal, oil, and—in England—electricity. The United Kingdom is the only country where electric heating, popular in the early 1970s as a room-based system, has lost its market share to natural gas. The substitution occurred mainly as a consequence of increasing central heating saturation.

Wood gained importance as a ready substitute for oil in regions where wood was plentiful and oil had been popular (Sweden, Norway, the northeastern United States, and parts of France). Roughly one-third of the wood used in these countries was used as a principal heating fuel, and the rest was used in conjunction with oil and electricity. Although uncertainties arise in counting the delivered or useful energy provided by wood, we judged that omitting wood from our analysis created a major bias. This is because so much wood use is found in homes otherwise using oil or electricity. Failure to count this wood gives a misleading picture of how the complementary fuels are used. In Sweden, for example, one-third of the homes using electricity or oil also used wood.

These changes were fostered by the availability of key household energy sources, which changed significantly during the years under observation: electricity became more available in France, natural gas in Italy, France, and Denmark. In Germany, by contrast, the availability of natural gas did not immediately boost its use because its price was pegged so closely to that of oil that oil users wondered whether they would save any money by abandoning their relatively new oil systems. Not surprisingly, the majority of homes situated where gas was available were still using other fuels for heating as late as 1985. In Britain, gas (where available) heats more than 80% of the homes, but future expansion of the gas system into new regions is expected to be slow. Thus, the pace of switching depends on availability and on relative prices.

To estimate the effect of fuel-switching, we calculated what the level of delivered energy demand would have been if only the mix of fuels had changed. This change reduced space heating demand in Europe by approximately 7%. In the case of the United States, the rise in electric heating in new homes reduced delivered energy by 8%, although the overall effect is clouded by the rise of wood as a secondary energy source, particularly after 1979. Given the universal use of multiple fuels for heating, this calculation is impossible for the case of Japan.

About 50% of the decline in oil use occurred through substitution, the rest through reduced intensities (Schipper and Ketoff 1985). We might consider most fuel-switching as a permanent component of change in energy use. In Denmark and Germany, for example, a family switching away from oil is required to dig up and dispose of the oil storage tank. However, in Sweden, Norway, and the northeastern United States, wood or electricity were used in place of oil, but the oil-fired equipment remained in place. Indeed, more than 2 million households in the United States switched from oil to wood in the late 1970s and early 1980s but kept their oil-burning equipment, perhaps awaiting the day when oil would be cheap again. Similarly in Japan, where more than 50% of homes combine the use of oil and electricity (in heat pumps), consumers can utilize the cheaper source more and the more expensive one less.

Changes in Intensity and Efficiency

Because space heating makes up such a large share of total residential demand, changes in space heating intensities had a significant impact on overall energy demand per household in almost every country we studied. The key measure of the intensity of heating is the ratio of useful energy for space heating to floor area.[6] This quantity can be normalized further by degree days to make comparisons between countries more accurate. Either of these indicators shows us what happened to space heating independently of changes in house size or other components of demand.

If all the structural components of space heating demand had remained unchanged, and only household energy use per square meter had varied since 1972, total energy consumed for space heating in 1987 would have dropped 37% in the United States, 18% in Japan, and almost 20% in Europe. Thus, decreases in unit consumption were great enough, for the most part, to offset structural changes that acted to increase total space heating demand.

Figure 8-9 shows the evolution of space heating energy intensity, normalized for differences and variations in climate and dwelling area. Denmark had the largest net decrease in intensity, while that in Norway increased. Changes in intensities can be attributed to energy conservation, both through short-term behavior modifications such as lower indoor temperatures and through long-term investments in measures such as insulation and more efficient furnaces. Long-term changes appear to have dominated in Sweden, where the change was gradual and nearly monotonic up to 1985. In most other European countries, short-term reductions were considerably larger and dominated the total changes. But these rapid changes occurred via behavior change, and are easily reversible. The rapid savings shown in Figure 8-9 for Denmark suggest a larger share of reversible savings induced by behavioral changes, a hypothesis confirmed by surveys monitoring household behavior in that country (Wilson et al. 1989).

By 1986, however, intensities began to increase in European countries, particularly in Germany and Denmark, where the earlier reductions were dominated by rapid, reversible behavioral changes. In the United Kingdom, growth in income stimulated a marked increase in the intensity of gas heating after 1984. By contrast, heating intensity remained virtually flat in the United States, where

6. Space heating intensities are presented in terms of useful energy in order to better compare the heating services delivered in countries with widely differing mixes of heating fuels.

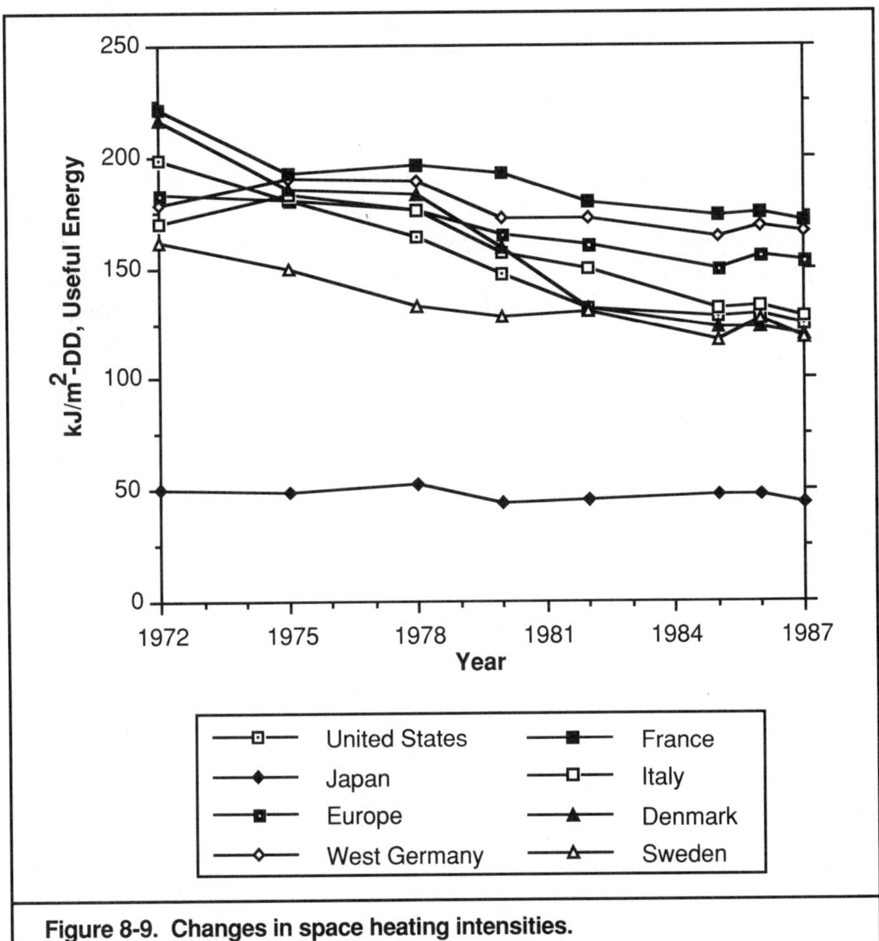

Figure 8-9. Changes in space heating intensities.
Note: DD = degree-day base 18°C (65.3°F); 1KJ = 0.948 Btu; 1 m^2 = 10.76 ft^2.

reductions in space heating intensity were more slow and gradual. In Norway and Japan, the decline in intensity after the first two oil shocks was reversed rapidly, revealing the high reversibility of changes in efficiency.

The permanency of changes in heating intensity is difficult to estimate. A detailed estimate would require disaggregated data not available for most countries, particularly data covering the last few years. Permanency of changes in intensity depends on buildings substitution rates, insulation investments, and changes in household behavior. The introduction of new, more efficient homes to the housing stock has certainly contributed to the increase in the permanent share of savings, and is likely to do so in the future, unless building thermal requirements established in most countries are relaxed. But the upturn in space heating intensities observed in 1986 and 1987 suggests that the reversible parts of intensity improvements are eroding rapidly.

The relative efficiency of new housing has contributed to the reduction in average heating intensity. Since 1973, thermal requirements for new homes were tightened in all countries, even in the country with the strictest levels, Sweden (Wilson et al. 1989). The increased thermal requirements for walls in new homes are shown in Figure 8-10 for six of the countries considered. This figure suggests that thermal requirements in most European countries were very lax in the early 1970s and only in 1984 reached the level established for Sweden in 1977.

The impact of these new, efficient buildings on average heating intensity is limited by the slow building rate in most of the countries

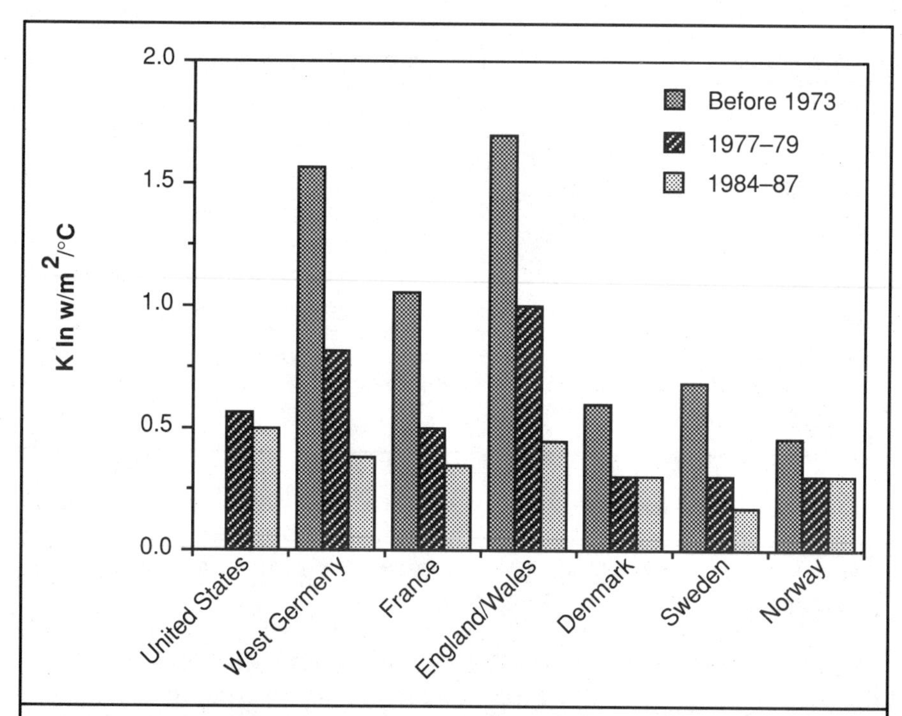

Figure 8-10. Thermal requirements in OECD countries: maximum heat transmission values for walls.

Note: "R" = 5.19 K.

analyzed here. Construction rates have been higher in the United States and Japan than in Europe, where the economic slowdown of the 1980s was accompanied by a reduction in household formation. Overall, 24% of the U.S. housing stock was built after 1975, while in Europe 15–18% was built after 1975 (EIA RECS; EIA AER). But in almost every country heating systems in new construction have been dominated by gas, electricity, or district heating. The impact of new construction on the average intensities of homes heated with these fuels was not insignificant; in fact, in the United States, Denmark, and France, as many as two-thirds of the homes heated by electricity in 1988 were built after 1975.

We can make a rough estimate of the impact of more efficient new construction in the following way. If homes built after 1975 comprise 15–20% of the 1987 stock and are 30% less heat intensive (in 1987) than the pre-1975 stock of homes in 1987, then average heating intensity among all homes in 1987 is about 5–7% lower than it would have been if new homes were no better insulated than those built before 1975. Since the overall intensity of heating, after consideration of the increase in central heating, fell far more than 5%, this means that the impact of efficiency improvements of new dwellings was only a small part of the overall reduction in heating consumption. This small impact on space heating of tightened thermal practices in new homes is permanent, and the effect on average household heating consumption will grow in time as more new homes enter the stock.

One important clue to the permanency of the decline in space heating intensity is the speed with which that decline occurred. The rapidity of the decline in the intensity of space heating varied from country to country, depending largely on the fuel mix. Oil intensities plummeted as much as 20% in two years, 1973 to 1975, in France, Germany, and Denmark, then rebounded through 1979, and fell as much as 25% through 1982. These drops could only have been caused by rapid changes in indoor temperature and other heating habits. We estimated that through 1983, at least 50% of the intensity decline in oil-heated homes in these countries (and in the United States) was reversible, and we have seen some upturn since 1983. Since oil was the most important heating fuel (except in the United States), the rapid changes in oil use dominated the change in total intensity. In France, Germany, and Denmark, as much as 40% of the decline in space heating intensity could be termed reversible. Although oil was a key fuel in Sweden, the decline in oil intensities there was slow and steady, indicating that most of the changes were technical and therefore not easily reversible.

By contrast, gas and electric heating intensities did not fall as rapidly. Moreover, the number of homes using gas or electricity increased rapidly because of new construction, and, in a few cases, conversions. New homes were more efficient than existing ones; the entry of these new homes into the stock had a slow and steady downward impact on average gas or electric heating intensity. The gradual reduction in heating intensities of gas and electric homes indicates that technical improvements, not sudden behavior changes, played a significant role in the conservation we observed. Thus in the United States, where gas is the most important heating fuel, the more gradual decline in intensity indicated that slow but steady technical improvements were more important than sudden reversible behavioral changes. We judge that in the United States only 25% of the savings in space heating could be easily reversed. And in Britain, where gas intensities have been steady or even rising during the past ten years, there is little conservation to reverse. Similarly, there has been little behavioral conservation in Norway, while in Japan rapid declines in kerosene use were followed by rebounds when prices receded.

The intercountry disparity in the space heating energy intensity was reduced over the study period. This reduction occurred in part because the reduction in unit consumption in the United States and Denmark, two countries with the highest energy intensities in 1972–1973, was more than 30%, as contrasted with smaller decreases, or indeed increases, in all other countries. Intensities in the United States, Denmark, and Sweden decreased far more than they did in other countries. This suggests that households in higher-income, more comfortable environments are prepared to cut back on heating use by a variety of means, while those in less comfortable homes may do less. As a result of these decreases in space heating intensity, and increases in floor area, the differences in unit consumption among countries was much smaller in 1986–1987 than in 1972–1973. This important reduction in differences contributed to the narrowing of total consumption per household that occurred over the period of observation.

In summary, changes in the structure of space heating caused increases in household energy demand in almost all the study countries. Changes in the intensity of heating reduced demand in most, but not all, of the study countries. By 1987 heating use was significantly above its 1972-1973 levels in Japan and Norway, was well below the 1972 level in the United States and Denmark, and was slightly under 1972 values in the other countries in the study. Figures 8-11 to 8-13 (pages 251, 252, and 253) present the counteracting impact of those changes on total residential energy heating demand in the United States, Japan, and Europe. In all, structural growth exerted significant upward pressure on demand. In the next section we consider the other important use we analyzed, electric appliances.

Electricity Use for Appliances

Appliance electricity use has been the most important component of growth in residential energy demand. Average growth rates for the 1972–1987 period varied between 2.4% per year in Denmark and 6.5% per year in Japan. Growth slowed between 1979 and 1984, but increased thereafter. Consumption for appliances showed little decline in absolute terms or in unit household intensity, although

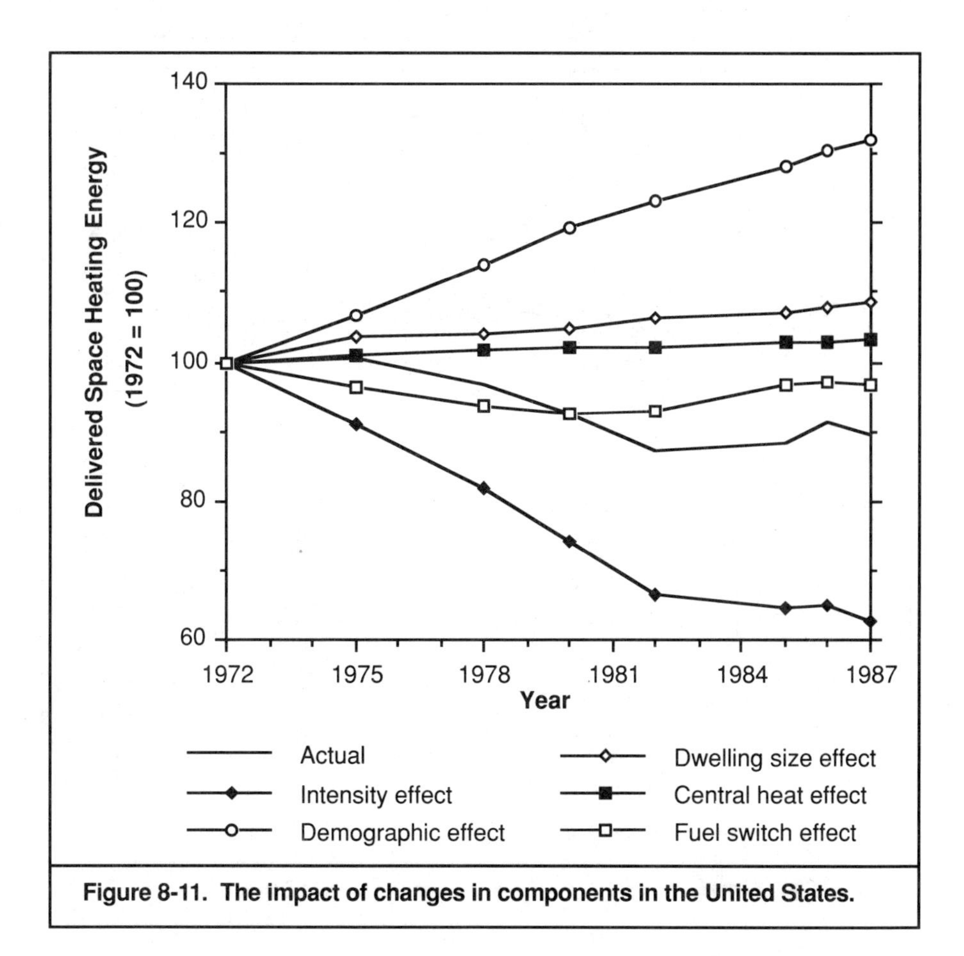

Figure 8-11. The impact of changes in components in the United States.

there were reductions in a few countries over limited periods (Schipper and Hawk 1989; Tyler and Schipper 1989).

Changes in electricity use for household appliances were driven by the increasing saturation, or penetration, of devices as well as by the changes in unit consumption for each appliance. Penetration or saturation refer to the share of households with a given appliance. Where we know the number of appliances per household, or diffusion, we give that figure. Diffusion is important for air-conditioning and refrigeration equipment. For these types of equipment, ownership of more than one device is important and tends to raise consumption per household.

The increased saturation, or penetration, of appliances exerted a vigorous upward pressure on electricity demand. Between 1972

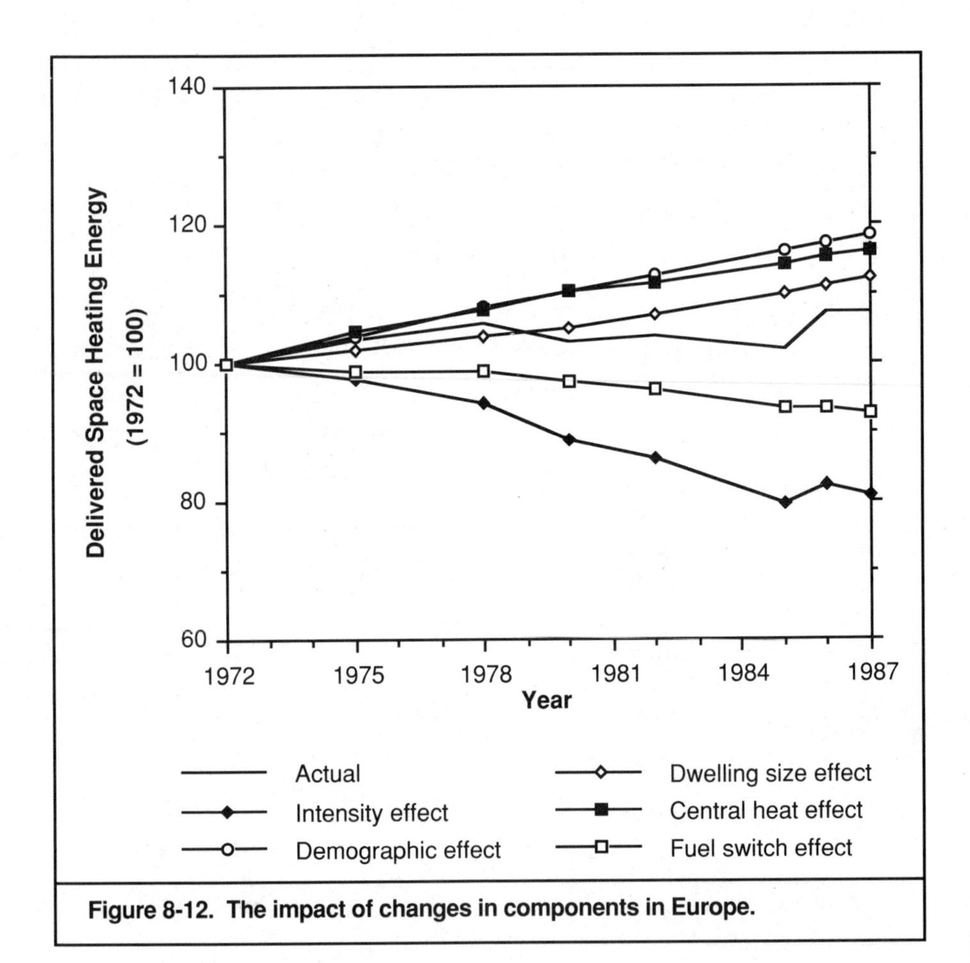

Figure 8-12. The impact of changes in components in Europe.

and 1987, the market penetration of washing machines, dishwashers, and freezers increased significantly in most countries, although the increases in the United States were less important than in the remaining countries, for whom ownership levels of most appliances were much lower in 1972, with the exception of refrigerator ownership, which was nearly saturated in all countries. However, the average size of a refrigerator increased steadily thereafter, and one-door models with small freezer compartments (or none at all) were replaced by two-door refrigerator-freezers.

Increased saturation of air-conditioning was significant in the United States and Japan. In the United States, central units gradually replaced room units. In Japan, room units spread rapidly. By the early 1980s, more than half of the new units sold in Japan also

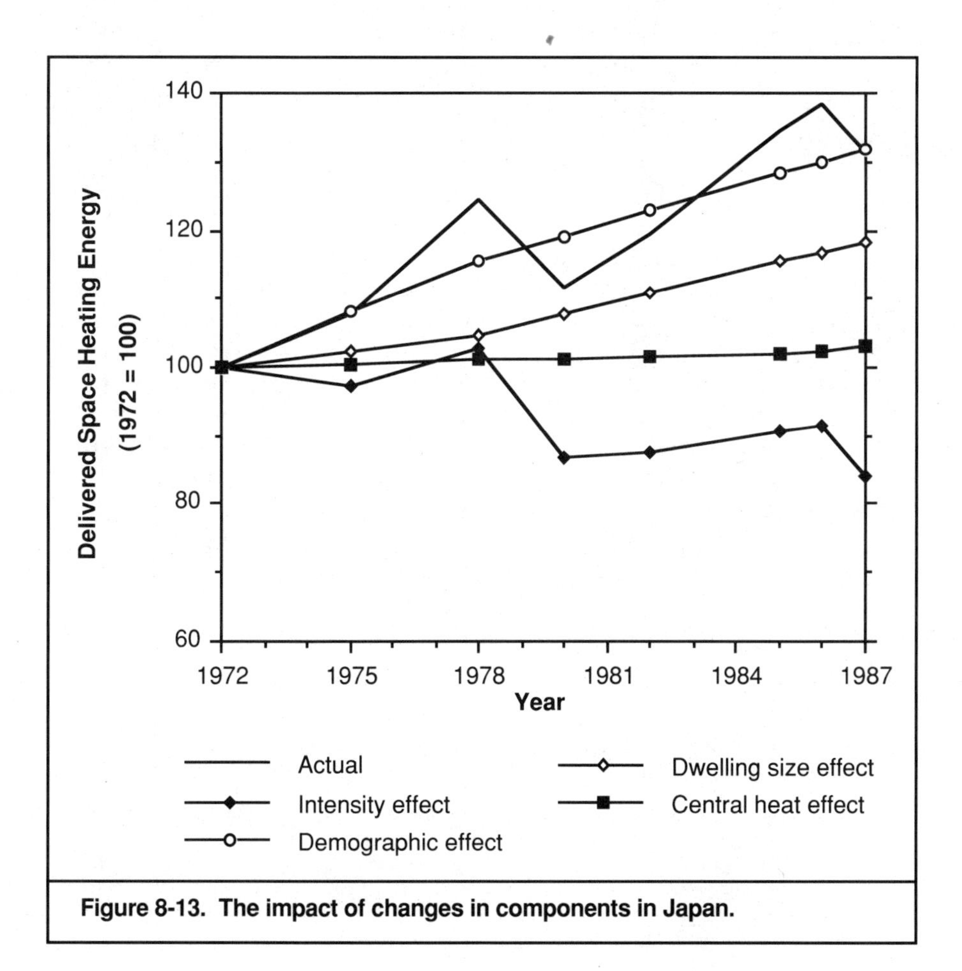

Figure 8-13. The impact of changes in components in Japan.

functioned as heat pumps during winter months. In European countries, by contrast, room air-conditioning was uncommon, but units are now becoming popular in Italy. In all, increased popularity of air-conditioning was important to the residential sector in the two largest OECD countries, and could become significant in southern Europe in the 1990s.

Ownership of dishwashers and clothes dryers is still increasing rapidly in several countries, including the United States. In France, for example, clothes dryers, virtually unknown in 1985, are now selling at well over half a million units per year (relative to 20 million households). Additionally, microwave ovens are selling well in all countries. While these are not major items, they substitute for electric or gas ovens.

We can estimate the impact of increased appliance ownership on total household electricity use. If the ownership of the five major appliances (refrigerator, freezer, clothes washer, dishwasher, and clothes dryer) had increased while unit consumption had remained the same, then European households would have used 28% more electricity for these uses than they did in 1972. For the United States, increased appliance ownership alone caused electricity use per household for major appliances to increase by 12% over 1972 consumption (20% if we include air-conditioning).

The electricity intensities (kWh per unit of service) of most electric appliances have fallen considerably since 1972, mainly as a consequence of improved technology. Figure 8-14 shows the index of electricity-use intensity for new refrigerators or refrigerator-freezers in three countries. (For other appliances the units of intensity or of its inverse, efficiency, differ.) Intensity (in kWh per liter of volume per year) is indexed to its 1975 value.[7] The changes in these three indices through 1985 are typical for all appliances. However, consumption did not fall as rapidly as did intensity, because increased use of some appliances, increased size, and a wider range of features exerted an upward pressure on unit consumption. Figure 8-15 (page 256) illustrates the shift in the size of refrigerators that occurred in Japan in the last 15 years (Japan Electric Manufacturers

7. The data for West Germany represent the consumption of a typical 210-liter (7.4-ft^3) refrigerator manufactured in the year shown. The data for Japan represent the consumption per liter for a large manufacturer's most popular (varying from year to year) combination refrigerator-freezers (hereafter called "combi"), which grew from under 150 liters capacity in the early 1970s to around 300 liters by 1987. The data for the United States represent the sales-weighted average consumption for all combis sold in a given year. Since refrigerators are tested differently in each country, absolute intensities are not directly comparable. For this reason, we present only an index of change.

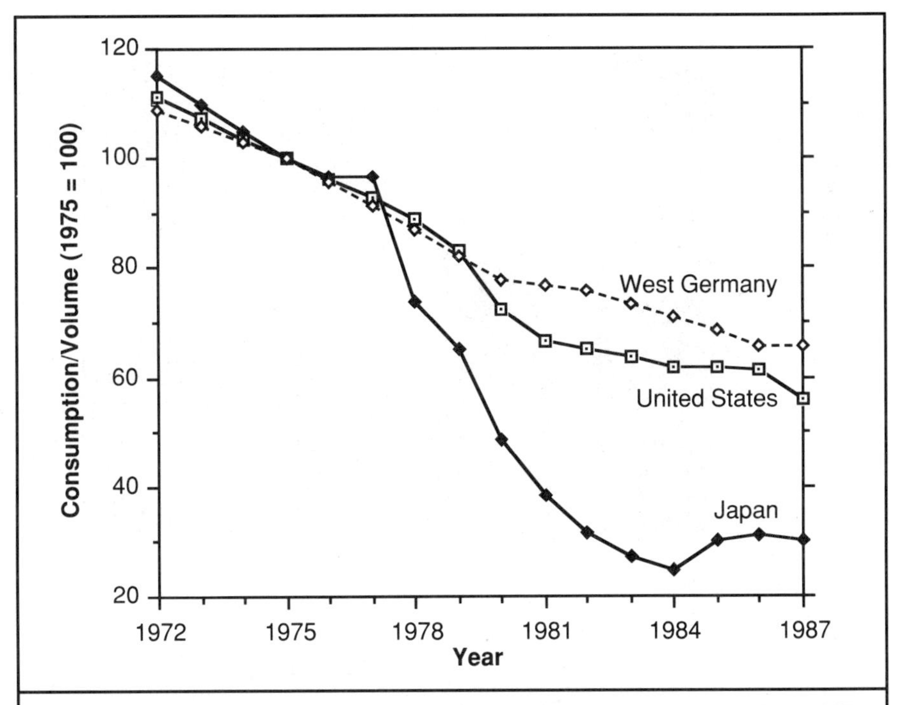

Figure 8-14. Refrigerator energy intensity: progress in new models, 1972–1987
Note: Different models are considered: West Germany = 1-door, 250 liter; United States = 2-door, 630 liter; Japan = 2-door, 150–330 liter (1 ft^3 = 28.3 lit.).

Association 1989). While the intensity of 150- to 300-liter refrigerators dropped almost 80% (Figure 8-14), these were largely replaced in the market by much bigger and therefore more energy intensive devices.

In general, unit consumption over the stock of all appliances fell. This decline occurred as old appliances were replaced with new and more efficient ones, and as newer, more efficient models dominated overall intensity. In a few cases, such as for washing machines, lower temperatures reduced unit consumption in existing machines significantly. (In our work, we exclude hot water from washing appliances to allow comparison between countries like the United States, where most of the hot water comes from the water heating system, and countries like Italy, where most appliances heat their own hot water.)

The overall impact on household electricity use from changes in appliances varied along the same lines as it did from changes in

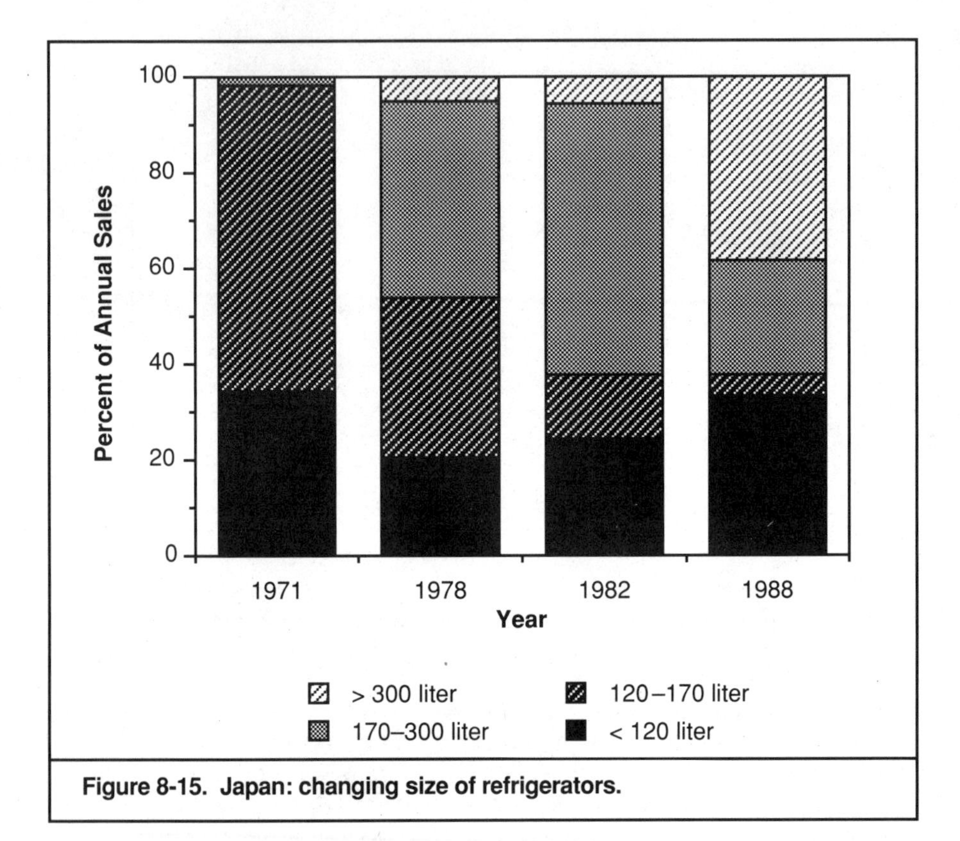

Figure 8-15. Japan: changing size of refrigerators.

space heating. In high-income countries, such as Sweden, Denmark, or the United States, appliance electricity use per household grew slowly. The impact of increased efficiency was sufficient to reduce or offset growth in electricity use per household generated by the increased appliance ownership. In Denmark, for example, this weighted average reduction in unit consumption of appliances was 23% (we estimate the impact of reduced intensities on household electricity use by evaluating the evolution of consumption per household if only unit consumption had varied). For the United States, the decrease was 17%. In other countries, growth in ownership and size of appliances was more important than improvements in efficiency, so electricity use increased. In Italy, for example, unit consumption as well as saturation both increased, pushing upward electricity use for appliances. Improvements in efficiency did occur, but increased size compensated for these improvements (Schipper and Hawk 1989; Tyler and Schipper 1989). In Italy, appliance electricity use grew both because saturation increased and because many important appliances were larger in 1986 than in 1972. This phenomenon is equivalent to the space heating transition from stoves to central heating.

Recently, however, efficiency improvements for appliances have slowed or even reversed. Increasing the number and extent of features of the refrigerator has increased the energy intensity. For example, refrigerator-freezers with fully automatic defrost or ice-makers use more electricity than those without. Thus, many factors besides efficiency influence the unit consumption of an appliance. In Japan, intensity is also rising because of the addition of a third or even fourth compartment and door, and due to the rising popularity of ice-makers. In Germany, the reduction in intensity has reached a plateau. In the United States, a similar plateau was reached, but the intensity resumed its decline in 1986 in anticipation of new efficiency standards implemented in California and soon in the United States as a whole (see Figure 8-14).

In all, more efficient appliances had a significant downward impact on household energy and electricity use. Electricity for appliances accounted for less than 15% of delivered energy to homes in 1972 (or roughly 35% of primary energy), but because household electricity use for appliances was rising at least twice as rapidly as fuel use for space heating, the importance of the increase in efficiency was magnified.

The Implications of Looking Beyond the Aggregate Picture

The People Who Use Energy: Americans in an International Perspective

Energy conservation in homes is about people and how they use energy. How have U.S. consumers done vis-à-vis their friends in Japan and Europe? We have shown that in the United States and most of the other countries studied, consumers have made substantial reductions in the intensity of household energy use since 1973. Most of the conservation in household energy demand occurred in space heating; this assertion is in part by assumption, but the large share of fuel use taken by space heating, and the large changes, imply that space heating changed more than water heating. Additionally, some of the reductions in intensity arose out of improvements in the efficiency of new electric appliances. These reductions were sufficient to offset most or all of the growth in energy demand that resulted from structural changes driven by demography and higher incomes. In this section we compare the overall performance of the United States with that of the other countries, and draw conclusions for future household energy use.

The reduction in space heating intensity in the United States, corrected for structural change, was slightly greater than the average for Europe. The reduction in Japan, by contrast, was smaller. The level of heating comfort in central Europe increased toward that of Sweden or the United States; the level in the United Kingdom, the lowest in the European countries, also increased after 1970 and closed in on standards in France and Germany. The level of heating in Japan reached that of the United Kingdom in 1970. These improvements in indoor comfort offset the impact of lower intensities on total energy use, leaving the impression that little conservation took place.

We want to point out that not all the space heating savings we measured were permanent. Savings were more rapidly introduced in oil-heated homes, suggesting that with time and lower price, these savings, mostly caused by changes in behavior, may in part wear off. Because of the limited importance of oil-heated housing stock, the portion of reversible savings is indeed smaller in the United States than in most other countries. Sweden stood out as a largely oil-heated country where savings were introduced slowly, principally through technology, and therefore least likely to reverse. Overall, the portion of residential heat savings in the OECD we deem reversible in 1987 is only around 33% of the difference between the 1987 intensities and the 1972 intensities.

The results we have reported here have important implications for future trends in demand for space heating. If structural growth has offset conservation achievements in the past, and if that growth is slowing because of saturation, then future conservation actions may decrease heating demand more than such actions have in the past. If the slack in prices of the late 1980s leads to a reversal of more of the savings in space heating, and if the pace of investments in improvements to existing and new homes also falls off, then household energy use for heating will grow at roughly the rate of increase in total heated area.

When we examine the overall changes in energy use for space heating that occurred, we are struck by how heating patterns in the United States and Europe converged. European households raised their heating comfort levels toward those of U.S. households. U.S. citizens lowered their heating intensity per household to that of the Europeans. Are there still differences between the structure of space heating in the United States and in other countries? A key difference remains: U.S. homes are still 40–60% larger than those in Europe and Japan, and the gap is only closing very slowly. As a result, per capita energy use for space heating in the United States has remained considerably larger than that in Europe. And the level of heating in Japan is still far behind that in the United States or Europe. This suggests that energy use for heating will increase for a long time in Japan, as long as prices do not shoot up or income growth does not fail.

What do the comparisons say about differences in efficiency? Homes in Sweden still have the most effective levels of insulation anywhere in the industrialized countries. Many homes in Norway and Denmark approach these levels. By contrast, the levels of insulation in the rest of Europe are much lower. These differences are seen if we look only at homes heated with oil, gas, or electricity; the differences are real, and not a result of our convention on aggregation (Schipper and Ketoff 1985; Ketoff et al. 1987; Schipper et al. 1987). From these differences we conclude that the United States could learn from construction and heating techniques practiced in Scandinavia. But the evolution of technology for building shells to date has been very much a local affair, so efforts to transfer technologies must pay close attention to the local conditions that influenced how these technologies were developed and implemented.

Officials of every country in the study were concerned with reducing oil use in the residential sector. Except for Germany and Japan, the consumption of oil in residential energy use was cut 33–75%, per capita use even more. In Japan there was simply no alternative, while in Germany the main alternative, natural gas, was expensive. In all countries except Japan oil heating was chosen in less than one-third of new housing starts, and it almost disappeared in Sweden and Norway. Notably, however, a majority of homes in the latter two countries still maintain the capability to use oil, so oil, while down significantly, is certainly not out. The use of other fuels increased, of course, in substitution for oil.

The story of household appliances is different for many reasons. First, almost all energy savings from appliances came through replacement, a slow process. Savings, if any, are permanent. Second, the supply of appliances is international; while there are local marketing, purchasing, and use differences, the technologies are similar everywhere. Thus, differences in performance may represent local preferences or simply availability of a particular technology, not a true local advantage in technology. This means that local conservation programs might improve the mix of what is offered for sale and actually sold, causing energy savings even before technical improvements are put in place. But the international nature of manufacturers means that efforts to improve technologies should be international, and coordinated with efforts to improve marketing and sales of efficient appliances. Few manufacturers want to produce a single product line for each country.

Other features of the evolution of appliance electricity use bear striking resemblance to the evolution of space heating. Electricity use for appliances was influenced by the same structural forces that shaped space heating. Growth in appliance ownership, and in their size and features, was far more important in Europe and Japan than in the United States. Households in Japan and Europe narrowed the gap in appliance ownership with the United States, but a major gap remains, particularly for freezers and refrigerators, where United States appliances are significantly larger than those in Europe or Japan.

Savings in electric appliances, measured as percentage reductions in the intensity of new appliances over time, are roughly the same order of magnitude as the reduction in heating intensity of new housing over the pre-1973 stock. Unlike space heating, differences in efficiency of appliances between the United States and other countries persist: in general, European and Japanese appliances use 20–33% less electricity, per unit of service, than do U.S. appliances.

Overall Impact of Conservation

What was the overall impact of energy savings? This impact is measured by multiplying the impact of the decline in energy intensity for space heating between 1972 and 1987 (see Figures 8-11 to 8-13) by the structural increases (more dwellings, larger area, and more central heating) of 1987, country by country. The same calculation is repeated for electric appliances. The results (Table 8-2) show

Table 8-2. Primary residential energy demand—actual demand and 1972–1987 savings (PJ).

	Actual demand		1972–1987 Savings			
	1972	1987	Space Heating	Electric Appliances	Total	Percent of 1987 Demand
United States	15,139	17,997	3,600	1,000	4,600	25.6
Japan	1,255	2,377	38	200	238	10.0
Europe[a]	8,262	10,449	1,250	375	1,625	15.6

a. Aggregate values for seven countries

how energy was saved for these two important end uses; that is, how much more primary energy would have been consumed had there been no reduction caused by conservation.

Note that in each of these regions, the actual 1987 primary energy use was higher than it was in 1972. Our disaggregated approach thus measured a significant amount of energy savings that are hidden with aggregate measures of changes in energy demand. These estimates explain the title of this project: only by disaggregating household energy demand into its structural and intensity components and analyzing the forces driving each component can we measure the impact of energy conservation on total household energy demand.

Impact of Policies

How much did energy conservation policies contribute to the savings we have measured? Measuring the impact of energy saving policies is very difficult and beyond the scope of our present work. However, we note several important observations:

1. The rapid drop in heating intensity after 1973 and again after 1979 (Figure 8-9) led to savings before new policies were put in place. Most near-term impacts appear to have been used by higher prices.
2. Thermal performance standards for new homes had a clear impact on construction techniques (Figure 8-10), but these could only have influenced approximately 75% of the stock built after 1973, or roughly 15% of the 1987 housing stock. The marginal impact of standards was certainly important, but could not have caused all the improvement in thermal performance.
3. Electric appliance standards in several U.S. states and informal agreements between government and manufacturers in Japan and Germany accelerated the improvement of efficiency of electric appliances between 1975/1976 and 1985 (Figure 8-14). However, it is difficult to say what would have happened to the efficiency of appliances without these policies.
4. Retrofit subsidies and loans reached at least 25% of households in Denmark and Sweden, and probably between 10% and 20% of households in the United States, Germany, and France (Wilson et al. 1989; EIA RECS). How many of these households would have undertaken conservation measures without subsidies has never been determined. Nevertheless, it is clear that a majority of households in every country changed heating habits without direct financial assistance from local or national governments.

What these observations indicate is that it is difficult to assign even half of the savings in household energy use to energy conservation policies promulgated in the countries we have studied. The observed changes in consumption were far greater than what could have been expected from the limited number of policy measures implemented, and the observed changes often occurred before policies were even put in place. Our best estimate is that through 1987, policies alone were responsible for 25–33% of household energy savings in the countries studied. This fact does not mean that policies were either ineffective or inefficient, only that other forces, such as changes in absolute or relative energy prices, caused more change than did policies.

Our observations do shed some light on the importance of energy prices. In France, Italy, Sweden, and Denmark, where heavy taxes on oil were imposed (accounting for more than 50% of the oil price) (IEA 1990), the greatest reductions in per capita oil use for homes were observed. In Germany and Japan, where virtually no taxes on energy or oil were imposed, the changes in oil use were far less significant. At the same time, the drop in prices, even in countries where taxes were significant, did lead to a slowing of the improvement in energy intensity, or even a rebound.

Implications for the Future

The political pressures for saving oil have receded in the late 1980s, but they have been replaced by strong interest in reducing the environmental impacts of energy use. Fossil-fuel use has been linked to increased CO_2 in the earth's atmosphere, leading to concern that climate will change more rapidly than at any other time in modern history. More efficient energy use reduces many of these problems associated with emissions. Can the rate of improvement in efficiency implied by the decreasing intensities we have measured be sustained? Can effective policies be designed—or redesigned—in the present era of uncertain energy prices?

These questions are particularly relevant in light of the observed plateau of the improvement in the efficiency of space heating (Figure 8-9). This slowdown may force reconsideration of demand-side energy savings goals, or at least require innovative programs or other mechanisms if space-heating intensities are to continue to fall.

We have seen evidence in some countries of short time horizons among consumers. The rapid changes in heating intensity in response to price changes suggest consumers are sensitive to prices in the short run. But do consumers make long-term decisions to invest in conservation? Unfortunately, there are few data on actual levels of insulation or other indicators of efficiency that would tell us what level of efficiency consumers, builders, or manufacturers really choose.

This lack of information presents a serious problem as nations approach the conference table to discuss reductions in emissions associated with fossil fuels. Certainly it is important to have some sense for the range of energy saving options that reduce emissions and save money, in order to make some broad ranking of the least cost options. But it is equally important to link observations of changes in energy use to policies that were put in place to encourage savings.

Unfortunately, this link is not apparent in most countries in Europe. Until recently many European officials did not even recognize the residential sector in official energy balances (Schipper et al. 1985; Schipper et al. 1990). Changes in the household sector are better documented in the United States. While the U.S. State Energy Data System is not exact, the figures published therein, combined with those from the Energy Information Administration's *Residential Energy Consumption Survey,* provide a useful picture of the evolution of the household sector since the early 1970s.

With little information on the structure of household energy use of the type we have presented in this chapter and in our previous studies, officials and analysts have been unable to separate the effects of energy conservation from other changes in energy use, unable to pinpoint the savings caused by various technologies, and unable to relate the changes observed to those in the housing or equipment stock as a whole. Authorities therefore could not isolate the savings that might have been caused by policies. How can we plan policies for the possibility that environmental forces will dictate an accelerated tempo of household energy savings if we are unsure of what policies have accomplished or of which policies and techniques were most effective in the past? In other words, what are the real economics of past and future energy savings?

Acknowledgements

The authors would like to acknowledge the support of many individuals in the United States and abroad who provided important sources of information for this study. We also thank Jeff Schlegel, Ed Vine, and Omar Masera for their thorough reviews.

This work was supported by the Assistant Secretary of Conservation and Renewable Energy, Office of Building and Community Systems, of the U.S. Department of Energy under contract No.

DE-AC03-76SF00098, and by the Dipartimento di Energetica ed Applicazioni di Fisica, Universita` degli Studi di Palermo, Italy, under contract BG 89-148.

References

Carlsson, L. 1989. *Energianvaendning och strukturomvandling i byggnader 1970–1985*. R22:1989. Stockholm: Byggforskningsraadet.

EIA. See Energy Information Administration.

Energy Conservation Center. 1989. *Energy Conservation in Japan*. Tokyo: MITI.

Energy Information Administration. Various years. *Annual Energy Review* (AER). Washington, D.C.: U.S. Department of Energy.

——. Various years. *Residential Energy Consumption Survey (RECS)*. Washington, D.C.: U.S. Department of Energy.

IEA. See International Energy Agency.

International Energy Agency. 1987a. *Energy Balances of OECD Countries, 1970–1985*. Paris: OECD.

——. 1987b. *Energy Conservation in IEA Countries*. Paris: OECD.

——. 1989a. *Energy Balances of OECD Countries, 1986–7*. Paris: OECD.

——. 1989b. *Electricity End-Use Efficiency*. Paris: OECD.

——. 1990. *Energy Prices and Taxes*. Paris: OECD.

Japan Electric Manufacturers Association. 1989. *Annual Report*. Tokyo.

Ketoff, A., S. Bartlett, D. Hawk, and S. Meyers. 1987. "History and Prospects of Residential Energy Demand in OECD Countries." Paris: Energy Demand Symposium, International Energy Agency.

Ketoff, A., and L. Schipper. 1990. *Household Energy Conservation: The Components*. LBL-28768. Berkeley, Calif.: Lawrence Berkeley Laboratory.

Schipper, L. 1987. "Energy Conservation Policies in the OECD: Did They Make a Difference?" *Energy Policy* 15 (10).

Schipper, L., and A. Ketoff. 1983. "Home Energy Use in Nine OECD Countries." *Energy Policy* 11 (2).

Schipper, L., A. Ketoff, and A. Kahane. 1985. "Explaining Residential Energy Use by International Bottom-up Comparisons." *Annual Review of Energy* 10.

Schipper, L., and A. Ketoff. 1985. "The International Decline in Household Oil Use." *Science* 230.

——. 1986. "The Rise and Fall of the Residential Oil Market: The Role of Consumers." In *Consumer Behavior and Energy Policy—An International Perspective,* ed. E. Monnier et al. New York: Praeger.

Schipper, L., A. Ketoff, S. Meyers, and D. Hawk. 1987. "Residential Electricity Consumption in Industrialized Countries: Changes Since 1973." *Energy* 12 (12).

Schipper, L., and D. Hawk. 1989. *More Efficient Household Electricity Use: An International Perspective*. LBL-27277. Berkeley, Calif.: Lawrence Berkeley Laboratory.

Schipper, L., S. Bartlett, D. Hawk, and E. Vine. 1989. "Linking Life-styles and Energy Use: A Matter of Time." *Annual Review of Energy* 14.

Schipper, L., R. Howarth, and H. Geller. 1990. "United States Energy Use from 1973 to 1987: The Impacts of Improved Efficiency." *Annual Review of Energy* 15.

Tyler, S., and L. Schipper. 1989. *Residential Electricity Use: A Scandinavian Comparison*. LBL-27276. Berkeley, Calif.: Lawrence Berkeley Laboratory.

Wilson, D., L. Schipper, S. Tyler, and S. Bartlett. 1989. *Policies and Programs for Promoting Energy Conservation in the Residential Sector: Lessons from Five OECD Countries*. LBL-27289. Berkeley, Calif.: Lawrence Berkeley Laboratory.

Index